Evaluation Platform of Sustainability for Global Systems

Aki-Hiro Sato · Hiroe Tsubaki

Evaluation Platform of Sustainability for Global Systems

Statistical Approach to Geospatial Information for Big Data Integration

Springer

Aki-Hiro Sato
Department of Data Science, Graduate School of Data Science
Yokohama City University
Yokohama, Kanagawa, Japan

Hiroe Tsubaki
The Institute of Statistical Mathematics
Tachikawa, Tokyo, Japan

ISBN 978-981-97-2295-2 ISBN 978-981-97-2296-9 (eBook)
https://doi.org/10.1007/978-981-97-2296-9

This Springer imprint is published by the registered company Springer Nature Singapore Pte Ltd.
The registered company address is: 152 Beach Road, #21-01/04 Gateway East, Singapore 189721, Singapore

Bonum commune naturae, communitati, et hominum
A common good for nature, community, and individuals

Preface

This book provides concepts to evaluate sustainability in global systems from theoretical and methodological perspectives.

Sustainability is essential for considering our society's persistence for the next generation. Almost all sustainability issues are formalized as multi-dimensional optimization problems consisting of independent holistic objective functions because many separate objective functions are balanced coincidentally for sustainability problems.

Moreover, people must share key performance indicators to solve multi-dimensional optimization problems related to such sustainable issues. However, many sustainable issues are challenges in part of global systems, modeled as complex systems with many components. Social, economic, and environmental aspects in our socioeconomic systems need to design solutions for the sustainability of our society.

This implies that Big data must contribute to evaluating our current situation, analyzing gaps between desirable and current situations, designing plans, and ensuring actions based on continuous improvement concepts, well-known as a Plan-do-check action (PDCA) cycle.

To do so, we must know how to construct a platform using Big data for continuous improvement. Thus, extracting meaningful information from Big data is one of the crucial issues in the current advanced information society.

Computational complexity for Big data is vital in measuring and quantifying system characteristics. It is a challenge to solve and requires enormous computational resources for computing Big data.

Primarily, this book considers spatio-temporal aspects of Big data issues and adopts Grid Square statistics as one of the efficient solutions for approaching geospatial data for Big data integration since time and space are unique axes in real-world problems.

Various types of Grid Square statistics are produced in different countries under their Grid coordinate system. To exchange Grid Square statistics for business and government purposes, we are also considering developing international standards for Grid Square statistics.

This book proposes the World Grid Square Data Reference Architecture for constructing a distributed data platform that implements and deploys Web data applications across organizations and domains. We explain and design the concept, system architecture, distributed data storage, parallel computation, and implication of Web data applications under the proposed distributed architecture. As a case study, we show our test-bed platform, MESHSTATS, a distributed data analytics platform, from both standardization and constructionism points of view.

This book also shows some case studies that evaluate socioeconomic systems based on Grid Square statistics and their applications to socio-economic environmental issues such as tourism, an SDGs indicator, natural disaster assessment, and spatial natural language processing.

Grid Square statistics is a valuable method to measure socio-economic-environmental systems holistically and provide a fundamental data infrastructure to optimize and monitor our socioeconomic activities. From this book, readers will be able to learn how to use Grid Square statistics and employ distributed architecture for Grid Square statistics and their applications.

Yokohama, and Kyoto, Japan — Aki-Hiro Sato
Tokyo, Japan — Hiroe Tsubaki
January 2024

Acknowledgements

This book contains outcomes of the research project financially supported by JST MIRAI entitled "Architecture design and actual proof for self-distributed World Grid Square statistics platform." (Project leader: Prof. Dr. Aki-Hiro Sato; Grant Number: 2020 JPMJMI20B6; Duration: November 2020–March 2023). The purpose of the JST MIRAI project is to clarify the ideal architecture of an integrated Grid Square statistics infrastructure that enables us to understand and explain the situation of socio-economic-environmental systems, predict future projections, estimate unobservable socioeconomic variables, and realize automatic decision-making. Thus, the goal of the JST MIRAI project was to clarify the architecture for the self-distributed World Grid Square statistics platform and demonstrate a prototype in an actual environment from both organizational and technical aspects.

The JST MIRAI project attempted to

- develop highly self-distributed data platform based on World Grid Square statistics and data,
- balance between privacy and public utility for data utilization and AI solutions, and
- clarify system architecture and its requirements of self-distributed World Grid Square statistics platform from both organizational and technical aspects.

We explored sustainable implementation for data utilization under the developed architecture through the JST MIRAI project.

Specifically, the authors would like to express their sincere gratitude to Prof. Dr. Manabu Iwasaki (The Institute of Statistical Mathematics), Mr. Shigehiro Kato (Recruit Co., Ltd.), Mr. Norihiro Suganami (Tanseisha Co., Ltd.), Mr. Syoki Nishimura (MIC and National Statistics Center), and Prof. Dr. Tomoko Matsui (The Institute of Statistical Mathematics) for stimulative discussion through the JST MIRAI project as research project members. The authors are also thankful to Mr. Kazutoshi Maki (GENEX Partners), Dr. Meiko Jensen (Karlstad University, Sweden), Prof. Dr. Mikio Suga (Hosei University), and Mr. Shinichi Nagao (Statistical Research and Training Institute) for valuable advice through the JST MIRAI research project as advisors.

The authors were further encouraged by Mr. Shinji Matsui (MIC), Mr. Naoki Makita (MIC), Dr. Naofumi Sakaguchi (MIC), Prof. Fumihiko Nishi (Yokohama City University), Dr. Koya Sato (Incremental, Inc.), and Prof. Dr. Tatsuto Asakawa (Waseda University), Ms. Kazumi Wada (MIC), Dr. Toshiyuki Shimono (Research Institute for World Grid Squares), and Mr. Yusuke Kemmochi (Research Institute of Spatiotemporal Behavior Chains, Co., Ltd.) during the discussion about Grid Square statistics in the JST MIRAI project.

One of the authors (Aki-Hiro Sato) are also thankful to Ms. Mie Yamato and Ms. Yumi Mizumura for their warm support in maintaining the research environment, to Mr. Desmond Fomo for preparing the manuscript by reviewing the draft, and to Ms. Su Xuanrong for preparing the manuscript by drawing some figures.

Moreover, this book includes part of the research outcomes conducted by a research project entitled "Evaluation Platform for Sustainability of Global Systems" (PRESTO Researcher: Dr. Aki-Hiro Sato; JST PRESTO Grant Number JPMJPR1504; October 2015–March 2020) is financially supported by the Japan Science and Technology Agency (JST).

The purpose of the JST PRESTO project is to establish fundamental Big Data technology for contributing to enhancing sustainability. UNEP has promoted sustainable production and consumption (SPC) in its 10YFP since 2013. There were five branches (consumer information, sustainable lifestyle and education, sustainable public procurement, sustainable building and construction, and sustainable tourism) in the 10YFP. Specifically, the JST PRESTO project focused on sustainable tourism in the SCP and attempted to construct an evaluation platform to monitor the state of the tourism sector based on data from finance, transport, and tourism. It developed a system to measure regional sustainability from several dimensions (economics, labor force, social structure, capacity, and energy and material consumption) regarding the interaction between the system and the real world. The JST PRESTO project attempted to develop data, a cloud-based platform, computational methods, and organization for monitoring sustainability in a data-driven manner.

Furthermore, some of the Grid Square data and statistics used in this book were computed using computational resources provided by the Institute for Mathematical Statistics based on an HPCI project (hp160060).

Contents

Acronyms

AMA	Administrative Management Agency
ANZLIC	Australia and New Zealand Land Information Council
API	Application Programing Interface
BDRA	Big Data Reference Architecture
BNG	British National Grid
bps	Bit per second
DGGS	Discrete Global Grid System
DMAIC	Define-Measure-Analyze-Improve-Control
DS	Data Science
EIPM	Evidence-Imformed Policy-Making
ETRS	European Tereserial Reference System
EU	European Union
GGRF	Global Geodetic Reference Frame
GPS	Global Positioning System
GSBPM	Generic Statistical Business Process Model
GSGF	Global Statistical Geospatial Framework
GUI	Graphical User Interface
HTTP	Hyper Text Transfer Protcol
ICT	Information Communication Technology
JIS	Japanese Industrial Standards
KFI	Key financial indicator
KGI	Key goal indicator
KPI	Key performance indicator
KSF	Key success factor
LAEA	Lambert Azimuthal Equal-Area Projection Coordinaate Reference System
NNG	National Nested Grids
P2P	Peer-to-peer
PDCA	Plan-Do-Check-Action
POI	Point-of-Interest
QC	Quality Control

REST	Representation State Transfer
SCC	Strongly connected components
SDGs	Sustainable Development Goals
TQC	Total Quality Control
TQM	Total Quality Management
UK	United Kingdom
UN-GGIM	The United Nations, Committee of Experts on Global Geospatial Information Management
UNSC	The United Nations Statistical Commission
UTM	Universal Transverse Mercator
WGS	World Geodsic System
WGSDRA	World Grid Square Data Reference Architecture
WWW	World Wide Web

Chapter 1
Introduction

Abstract This chapter depicts the background and concepts of a platform designed to evaluate global systems. Sustainability is a crucial concern within our socioeconomic-environmental systems and is formalized as multi-dimensional optimization problems containing several key performance indicators. Big data has enabled us to analyze and improve our world from a holistic point of view. Moreover, geospatial big data enhances the understanding of sustainability from regional aspects. However, they have several challenges from data and computation points of view. The utilization of Grid Square statistics can contribute to addressing critical issues to evaluate the sustainability of our world by enabling us to share and compute data and statistics in a reasonable way based on their ability on discretization or digitalization. Furthermore, this chapter addresses the purpose and motivation of this book and its structure.

1.1 Sustainable Issues and Our World

The 2030 Agenda for sustainable development adopted by the United Nations General Assembly in 2015 [1] issued Sustainable Development Goals (SDGs) consisting of 17 Goals and 169 associated Targets, measured by 231 indicators.

Sustainability is an essential issue for considering our society's persistence for the next generation. The paradigm of sustainable development is described on three conceptual pillars: *economic sustainability*, *social sustainability*, and *environmental sustainability* [2].

According to Kahn [3], three pillars of sustainability are described as follows:

- Economic sustainability is measured by growth, development, and productivity. The sustainable development theory expands monetary capital in a conventional development to natural, social, and human capital.
- Social sustainability is captured by equity, empowerment, accessibility, participation, sharing, cultural identity, and institutional sustainability.

A.-H. Sato and H. Tsubaki, *Evaluation Platform of Sustainability for Global Systems*,
https://doi.org/10.1007/978-981-97-2296-9_1

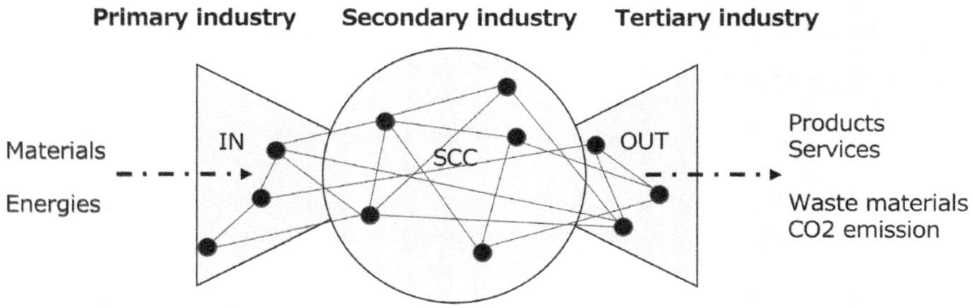

Fig. 1.1 Bow-tie structure consisting of IN, SCC, and OUT components as a network model of our society

- Environmental sustainability includes ecosystem integrity, carrying capacity, and biodiversity. The natural resources must be consumed slower than their regeneration speed since natural capital are as sources of economic inputs and a sink for waste.

Expressly, social, and economic developments conflict with environmental sustainability due to consuming natural resources of socioeconomic systems. Figure 1.1 shows a conceptual illustration of the bow-tie structure of the society. It models companies conveying materials in our society as a bow-tie structure consisting of input components (IN), strongly connected components (SCC), and output components (OUT) from a network viewpoint. Hearnshaw et al. conducted a complex network analysis for a global supply chain network [4]. Some studies on network analysis for the society show that components are characterized in industrial categories [5–7]. Nodes represent companies and organizations, and links express material exchanges. The IN obtains resources and energies from natural environment. The SCC produces artificial objects (products) from materials using energy. The OUT distributes products to consumers. Our society exhausts waste materials and CO_2 emissions as byproducts. Companies are placed in the IN, mainly of the primary industry, in the SCC of the secondary industry, and the OUT of the tertiary industry.

Moreover, most sustainability issues are formalized as multi-dimensional optimization problems because socioeconomic objective functions and environmental ones are independent but balanced coincidentally for sustainable issues. Moreover, people and companies need to share Key Performance Indicators (KPIs) to solve multi-dimensional optimization problems related to sustainable issues realized in our real-world [8–10].

Thus, we should measure sustainability from three components such as social, economic, and environmental. Our socioeconomic systems are generally constructed by mobility, social networks (relationships), and accessibility to information and resources. Mobility may create an opportunity to strengthen ties among people and can change social network structures. In addition, accessing information and resources may form the motivation structure of mobility.

The social network defines chance and necessity to give accessibility to information and resources. These problems need a multi-dimensional approach to socioeconomic systems.

Moreover, to change our socioeconomic systems, we need to transform mobility, social networks, and accessibility to information and resources. Social design methods [11, 12] and implementation procedures help realize sustainable concepts in our socioeconomic systems. How should we construct our world? The implementable social design needs knowledge about different sectors and organizations.

1.2 Popper's Three Worlds and Sustainability

Karl Popper proposes a concept of three worlds [13]. World I is a world of natural environments. Physical objects are its inhabitants. This world existed even before human beings appeared on the Planet. World II is a world of human recognition and psychology. Subjective experiences are its inhabitants. Our human brains have languages by expressing words described by characters and letters, which revealed from our psychological activities; World III is a world produced by our human activities inside World I. World III defines the subsistence of human beings and the range of our actions. We make artifacts from substances that consume energy obtained from World I and exhaust waste matters and CO_2 to World I.

The sustainable issues are understood as limits of World III constrained by World I. Therefore, the size of World III should be smaller than that of World I. Thus, almost issues of sustainability need reconsideration regarding the design of artifacts.

However, since the scale of our world is enormous, it is difficult for us to capture our world quantitatively. Moreover, since our human ability is limited, we need several technologies. Information communication technology (ICT) and data science (DS) may help us to improve our decision-making and find better solutions to sustainable issues.

Furthermore, such knowledge about multi-sectors is beyond the capability of one person. One of the most prominent way is a use of big data that is accumulated in various ICT systems and has not been used effectively, yet. Thus, we must use big data to understand the current situation of our society. The big data may further enable us to approach to our world from a holistic point of view.

1.3 Big Data

According to ISO/IEC 20546:2019 (Information technology-Big Data-Overview and vocabulary) [14], *big data is defined as extensive data that requires a scalable technology for efficient storage, manipulation, management, and analysis.*

Data can be preliminarily characterized by 4V (Volume, Velocity, Variety, and Variability). Data volume is the extent of the amount of data relevant to impacting computation and storage resources and their management during data processing. Data velocity is the rate of flow at which data is generated, transmitted, stored, analyzed, or visualized. Data variety is defined as a range of formats, logical models, timescales, and semantics of the dataset. Finally, data variability means temporal changes in the transmission rate, format, structure, semantics, or quality of datasets.

Big data are high volume, velocity, variety, and variability datasets. Indeed, big data is a relative concept between the system's data and the information processing capacity that handles the data. Thus, processing, handling, and analyzing big data require scalable technology.

To understand big data intuitively, we start with a comparison between human ability and big data produced within 2020. Suppose that the fundamental information processing unit is our human. From knowledge about the number of synapses in the cortex of human brains, the size of human memory is estimated to be 10^{13} bits [15] John Von Neumann estimates it to be 10^{20} bits [16] from the sum of all neural impulses generated in the brain throughout life. Moreover, human processing speed is limited due to the minimum impulse length of neuronal cells at 1 ms [17]. Therefore, there is the minimum recognizable time duration that is determined by an upper limit on the firing rate of synaptic cells.

From several experiments, human long-term memory speed is typically about 2 bits per second (bps) [15] in terms of reading and listening. If we assume that human life is about 80 years, so that 2.5 giga seconds (Gs) (giga means 10^9), then the maximum amount of long-term memory is 625 megabytes (MB) (2 bps $\times$ 2.5 Gs $= 5 \times 10^9$ bits), which is less than 0.6% of the physical size of memory in human brains. If we manually process more than 625 MB of text data, then we need to spend more time than the total lifetime. Therefore, since such data is bigger than our human information processing capacity, it could be called big data compared with human information capability.

The IDC's report [18] predicts, that the Global Datasphere will grow from 45 Zettabytes (ZB; Zetta means 10^{21}) in 2019 to 175 ZB by 2025. The digital data produced or replicated in 2025 will become about 10^{16} times larger than human long-term memory capacity per year (about 7.8 MB/year). Therefore, we could assume that this huge amount of digital data generated were stored but not used for some purpose. If we can find a way to use and analyze data that just have been kept, we may obtain various valuable insights for decision-making.

We further need a careful strategy to control data quality and reduce data bias in the case of big data since we have no way to check data quality and bias for big data manually. Moreover, data quality management for big data is highly costed and requires some platform with scalable technology.

It is known that exploratory data analysis [19] helps improve the quality of data analysis in general. Big data analytics needs continuous improvement process based on exploratory data analysis.

1.4 Data Analytics Pipeline

To introduce the multi-dimensional approach to social design, we establish a mutual understanding of different sectors and organizations. Data utilization is one of the most effective solutions. Roughly speaking, data analytics is accomplished by several steps such as data acquisition, data understanding, data preparation, data visualization and analysis, and data interpretation. After the data interpretation stage, the results should be reported as a specific document.

It is known that most data scientists spend only 20% of their time on actual data analysis and 80% of their time on data preparation including data acquisition and data understanding [20]. Therefore, data infrastructure should be essential to tackle the issue of the time-consuming problem of data preparation. The data infrastructure requires several functions, described as requirements such as data normalization, data insertion, data update, data search, data extraction, data calculation, and data deletion. In addition, standardized data normalization is essential to define the architecture of the data infrastructure.

First, we introduce the existing framework related to Grid Square statistics. Indeed, there are several existing frameworks related to data and statistics.

The United Nations General Assembly adopted a global geodetic reference frame (GGRF) [21] for sustainable development in 2015. The GGRF framework allows users to precisely determine and express locations on the earth based on standard methods and enables users to quantify changes in the earth in space and time.

According to the Global Statistical Geospatial Framework (GSGF) [22] adopted in the United Nations, Committee of Experts on Global Geospatial Information Management (UN-GGIM) and the United Nations Statistical Commission (UNSC), there are five principles:

- P1: Use of fundamental geospatial infrastructure and geocoding
- P2: Geocoded unit record data in a data management environment
- P3: Common geographies for dissemination of statistics
- P4: Interoperable data and metadata standards
- P5: Accessible and usable.

P1 means that fundamental infrastructure on geospatial data should be established to transform between addresses and geographical information. Standard definitions of latitude and longitude and a geodetic model are included.

P2 means that geospatial databases are maintained in a data management system.

P3 means a standardized definition of administrative units under common geographical for dissemination of statistics.

P4 means that data interoperability should be controlled in different sectors and organizations and that standards for metadata are defined.

Typically, interoperability can be described as a model with five facets: transport, syntactic, semantic data, behavioral, and policy. Table 1.1 shows five facets that are required for interoperability in cloud interoperability and portability as ISO/IEC

Table 1.1 Five facets required for interoperability

Facet	Aim	Objects	Solutions
Transport	Data transfer	Signals	Protocols of data, transfer
Syntactic	Understand format of transferred data	Data	Standardized data exchange formats
Semantic data	Interpretation of transferred data using a data model	Information	Common data models
Behavioral	Get anticipated outcomes when making service requests	Programmatic interface	UML models, pre-conditions, post-conditions, constraint specifications
Policy	Ensure that interacting systems conform to applicable laws, regulations and organizational policies	Laws, regulations, policies	Conditions for operating

19941:2017 [23]. The interoperability is fulfilled if and only if five types of different conditions are satisfied simultaneously.

P5 means the existence of a common data platform that users can access.

GSGF gives us a specific framework for geospatial statistics. The Generic Statistical Business Process Model (GSBPM) [24] is a workflow model for producing statistics differently within and between statistical offices. We can understand and share the status of producing data and statistics based on GSBPM as a common language.

Around 2006, Statistics New Zealand was the first organization to adopt the GSBPM in official statistics. Then, there was continuous work in joint UNECE/Eurostat/OECD Work Sessions on Statistical Metadata (METIS). The GSBPM has a hierarchical structure consisting of four levels:

- Level 0, the statistical business process;
- Level 1, the nine phases of the statistical business process;
- Level 2, the sub-processes within each phase;
- Level 3, a description of those sub-processes.

Level 1 of the GSBPM consists of nine phases: 1. Specify Needs, 2. Design, 3. Build, 4. Collect, 5. Process, 6. Analyze, 7. Disseminate, 8. Archive, and 9. Evaluate. The preparation is related to phases 1–3. The production ranges from phases 4–7. Phases 8 and 9 are actions for future improvement.

Table 1.2 Processes and sub-processes for preparation

Phase	Sub-process
1. Specify needs	1.1 Determine needs for information
	1.2 Consult and confirm needs
	1.3 Establish output objectives
	1.4 Identify concepts
	1.5 Check data availability
	1.6 Prepare the business case
2. Design	2.1 Design outputs
	2.2 Design variable descriptions
	2.3 Design data collection methodology
	2.4 Design frame and sample methodology
	2.5 Design statistical processing methodology
	2.6 Design production system and workflow
3. Build	3.1 Build data
	3.2 Build or enhance process components
	3.3 Configure workflows
	3.4 Test production system
	3.5 Test statistical business process
	3.6 Finalize production system

The Level 2 of the GSBPM are separated into each sub-process. Tables 1.2, 1.3 and 1.4 show processes and sub-processes for preparation, production, and future improvement. Phase 1 "Specify Needs", Phase 2 "Design", and Phase 3 "Build" consist of six sub-processes, respectively. Phase 4 "Collect", Phase 5 "Process", Phase 6 "Analyze", and Phase 7 "Disseminate" consists of four, eight, five, and five sub-processes, respectively. Phase 8 "Archive" and Phase 9 "Evaluate" consist of four and three sub-processes. The GSBPM is a framework for producing data and statistics for public and commercial uses through collaborative work by different persons.

1.5 Motivation and Structure of This Book

This book focuses on both theoretical and methodological aspects for and by an evaluation platform of sustainability for the global system. Firstly, we think of an evaluation as a central concept of data-driven manners and will explain methods and concepts of evaluation in Chap. 2. To evaluate and improve global systems continuously, an evaluation platform must have the following properties:

Table 1.3 Processes and sub-processes for production

Phase	Sub-process
4. Collect	4.1 Select sample
	4.2 Set up collection
	4.3 Run collection
	4.4 Finalize collection
5. Process	5.1 Integrate data
	5.2 Classify and code
	5.3 Review Variable and edit
	5.4 Impute
	5.5 Derive new variables and statistical units
	5.6 Calculate weights
	5.7 Calculate aggregates
	5.8 Finalize data files
6. Analyze	6.1 Prepare draft outputs
	6.2 Validate outputs
	6.3 Scrutinize and explain
	6.4 Apply disclosure control
	6.5 Finalize output
7. Disseminate	7.1 Update output systems
	7.2 Produce dissemination procedure
	7.3 Manage release of dissemination products
	7.4 Promote dissemination products
	7.5 Manage user support

Table 1.4 Processes and sub-processes for future improvement

Phase	Sub-process
8. Archive	8.1 Define archive rules
	8.2 Manage archive repository
	8.3 Preserve data associated metadata
	8.4 Dispose of data and associated metadata
9. Evaluate	9.1 Gather evaluation inputs
	9.2 Conduct evaluation
	9.3 Agree action plan

- Enable to collect, store, search, and compute data in a global scale (Big data issues)
- Enable to measure and evaluate global systems based on evidence for a long time (Sustainability)
- Enable to establish a man-machine system having communications with societies (Man-machine issues)

- Enable to operate and maintain a platform in resilient manner (Scalability, elasticity, and fault-tolerance)
- Enable to manage a platform in both a distributed and self-organized manner (Autonomous distributed system).

Therefore, we need to construct a platform with massive computer resources that balances actual demand from data and computation points of view. Spatio-temporal characteristics of data are essential to evaluate the status of global systems. However, it is challenging to realize geospatial big data integration effectively worldwide due to the limitation of several types of resources such as computation, knowledge, communication, social networks, and time.

Discretization is one of the solutions to handle such enormous geographical datasets since discretization of continuous space and time enables us to search data fast by a unique identifier code. This book introduces Grid Squares to identify unique data segments discretely, which will be explained in Chap. 3.

Grid Square statistics are geographical statistics produced using a tiny square area. Grids transform geographical information (latitude and longitude) into location IDs called Grid Square codes. Chapter 3 also explains the Japanese National Grid Square coding system and its extension for world usage.

Some countries adopt different grid coordination systems to produce Grid Square statistics. There are some national and international grid coordination systems.

Several types of Grid Squares have already been defined, and Grid Square statistics are produced based on different Grid coordinate systems in some countries and organizations. For example, the Japanese National Grid Square coding system is a national industrial standard registered as JIS X0410:2002 and World Grid Square coding system is a compatible extension to JIS X0410:2002 for worldwide usage. European Reference Grids is adopted in INSPIRE, EU Commission. Ordnance Survey National Grid is used in the United Kingdom. The ANZLIC's National Nested Grid is used in Australia, New Zealand, and Antarctica. Global Discrete Grid System (GDGS) is standardized internationally in ISO 19170-1:2021 [25].

Thus, we need standards to exchange Grid Square statistics among different coordination systems.

Grid Square statistics computed from data, including geographical information, provide one reasonable solution to handle such geospatial big data. We will mention Grid Square statistics in Chap. 4.

There is a high probability that high-resolution spatial data may easily violate the Protection of Personal Information and Personal Privacy Act. In addition, organizations having spatial data with high resolution should be careful when transferring their data to other organizations. Therefore, it is challenging to find the motivation to exchange high-resolution spatial data beyond different organizations and disciplines.

We attempt to introduce World Grid Square statistics produced from high-resolution spatial data as one of the solutions to exchange, analyze, and interpret data with geographical information. In that case, we may preserve the value of data and share data among different stakeholders without personal privacy issues. Moreover, we may create value-added data products by combining different Grid Square statistics from different data sources.

The typical architecture among different organizations enables us to integrate World Grid Square data and statistics from original data provided by various stakeholders and promote the development of dynamic data applications and data sharing.

Data integration beyond organizations enables us to realize data applications for decision-makers and monetize data by data harmonization. Therefore, the integrated data application platform based on World Grid Square statistics produced from different data sources may create high values. We will introduce how to construct data infrastructure for Grid Square statistics in Chap. 5.

We define a layered model for the system for approaching to geospatial big data. The layered model consists of five components: use cases, data applications, databases and algorithms, a platform, and computational resources. These five layers are abstractive components connected with the upper and bottom layers in each other. This layered model also follows international standard about big data reference architecture (BDRA) in ISO/IEC 20547-3:2020 [26].

Figure 1.2 shows a conceptual illustration of the five layers (use cases, data applications, databases and algorithms, a platform, and computational resources).

The use cases are practical and valuable usages of data applications in a physical environment. The data applications are computer software to visualize, calculate, quantify, and aggregate data based on some algorithms. Databases and algorithms are fundamental components to compute data with some logic and mathematics. The platform consists of some essential components for operating the data system. The

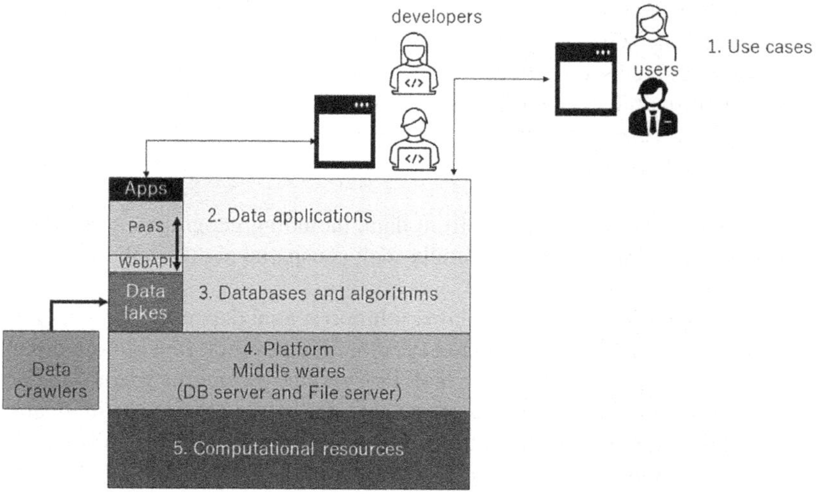

Fig. 1.2 The layered model for the system to integrate geospatial big data. The model consists of five components such as use cases, data applications, databases and algorithms, a platform, and computational resources

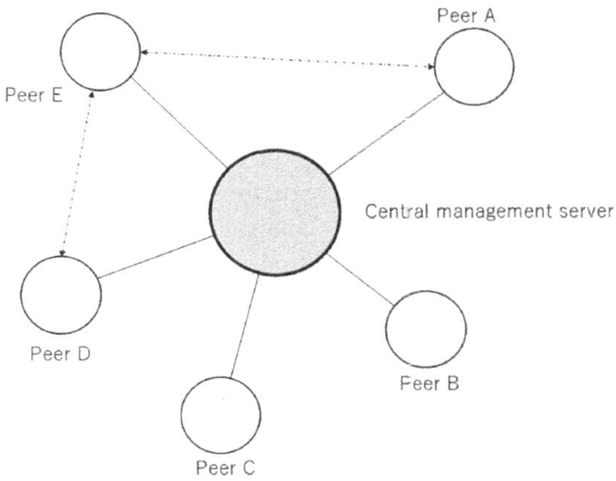

Fig. 1.3 A conceptual illustration of the peer-to-peer architecture. A Program can access one another via Internet

computer resources are physical and virtual computational equipment where some software runs.

We conclude that the distributed architecture of man-machine systems with organizational aspects under specific international standards should realize sustainability for a platform for evaluating global systems. Figure 1.3 shows an example of a peer-to-peer system with a super node. Platforms managed by different organizations that connect based on social agreements among organizations form such distributed autonomous systems.

To do so, different types of organizations need to provide a component for such a distributed system to evaluate global systems under expected types of social agreements such as international standards and regulations among countries. We will explain useful techniques in Chap. 6.

Moreover, we will introduce some case studies on Grid Square statistics and explain how to realize holistic integration using Grid Square statistics produced from data generated in different sectors and organizations. We will show case studies for producing Grid Square statistics in Chap. 7 and summarize some examples of data analysis in Chap. 8.

1.6 Conclusions

This chapter explained the importance of a data-driven approach to sustainable issues since sustainability requires continuous improvement for multi-dimensional optimization problems among economic, social, and environmental systems. Karl Popper's three worlds let us know the limit of growth of artifacts (World III) built in World I.

One of the most promising ways to approach sustainable issues is to promote the use of big data stored as digital data in data storage and may provide efficient solutions to maintain our socio-economic-environmental systems.

This chapter especially identified some processes to manage and integrate geospatial big data by introducing the Generic Statistical Business Process Model (GSBPM) to produce and use Grid Square statistics. Moreover, we defined a layered model to acquire, collect, process, analyze, and interpret for integration of geospatial big data under the physical or virtual computing resources. The distributed architecture is suitable for evaluating the sustainability of global systems in a data-driven manner for an extended period.

We further mentioned our motivation to write this book, which will explain handling, managing, and integrating geospatial big data using some Grid Square statistics techniques. Specifically, we elaborate to deal with theories and concepts that are useful at the time when we attempt to use Grid Square statistics to develop data applications for various stakeholders beyond different sectors and different organizations.

References

1. 2030 Agenda for Sustainable Development. https://sustainabledevelopment.un.org/post2015/transformingourworld. Accessed on 5 Jan 2023
2. A.D. Basiago, Economic, social, and environmental sustainability in development theory and urban planning practice. Environmentalist **19**, 145–161 (1999)
3. M.A. Kahn, Concepts, sustainable development: the key concepts, issues and implications, in Keynote paper given at the international sustainable development research conference, 27–29 Mar 1995, Manchester, UK, Sustainable Development, vol. 3, issue 2 (1995) pp. 63–69
4. E.J.S. Hearnshaw, M.M.J. Wilson, A complex network approach to supply chain network theory. Int. J. Oper. Prod. Manag. **33**(4), 442–469 (2013). https://doi.org/10.1108/01443571311307343
5. P. Crucitti, V. Latora, M. Marchiori, A topological analysis of the Italian electric power grid. Phys. A **338**(1-2), 92–97 (2004). https://doi.org/10.1016/j.physa.2004.02.029
6. S. Perera, M.G. Bell, M.C. Bliemer, Network science approach to modelling the topology and robustness of supply chain networks: a review and perspective. Appl. Netw. Sci. **2**(33) (2017). https://doi.org/10.1007/s41109-017-0053-0
7. A. Chakraborty, Y. Ikeda, Testing "efficient supply chain propositions" using topological characterization of the global supply chain network. PLoS ONE **15**(10), e0239669 (2020). https://doi.org/10.1371/journal.pone.0239669
8. S. Croom, P. Romano, M. Giannakis, Supply chain management: an analytical framework for critical literature review. Eur. J. Purch. Supply Manag. **6**, 67–83 (2000)
9. D. Power, Supply chain management integration and implementation: a literature review. Supply Chain. Manag.: Int. J. **10**(4), 252–263 (2005)
10. D. Helbing, *Managing Complexity: Insights, Concepts, Applications* (Springer, Berlin, 2008)
11. A.-H. Sato (ed.), *Applications of Data-Centric Science to Social Design: Qualitative and Quantitative Understanding of Collective Human Behavior* (Springer, Singapore, 2019)
12. I. Veiga. R. Almendra, Social design principles and practices, in *Design's Big Debates – DRS International Conference 2014* ed. by Y. Lim, K. Niedderer, J. Redström, E. Stolterman, A. Valtonen (eds.) (Umeå, Sweden, 2014), pp. 16–19. https://dl.designresearchsociety.org/drs-conference-papers/drs2014/researchpapers/42. Accessed 10 Jan 2024

13. K. Popper, Three Worlds, The Tanner Lecture on Human Values, The University of Michigan (1978)
14. ISO/IEC 20546:2019, Information technology—big data—overview and vocabulary. https://www.iso.org/standard/68305.html. Accessed 9 Oct 2022
15. T.K. Landauer, How much do people remember? some estimates of the quantity of learned information in long-term memory. Cogn. Sci. **10**(4), 477–493 (1986)
16. J. Von Neumann, *The Computer and the Brain* (Yale University Press, New Haven, CT, 1958)
17. G. Salvendy (ed.), *Handbook of Human Factors and Ergonomics*, 4th edn. (Wiley, Hoboken, NJ, 2012)
18. D. Reinsel, J. Gantz, J. Rydning, The digitization of the world from edge to core (2020), an IDC White paper - #US44413318, Sponsored by SEAGATE. https://www.seagate.com/files/www-content/our-story/trends/files/dataage-idc-report-final.pdf. Accessed 10 Jan 2024
19. J.W. Tukey, *Exploratory Data Analysis* (Addison-Wesley, Reading, MA. 1977)
20. CrowdFlower Data Science Report 2016. https://visit.figure-eight.com/rs/416-ZBE-142/images/CrowdFlower_DataScienceReport_2016.pdf. Accessed 5 Jan 2023
21. UNGGRF. https://www.unggrf.org. Accessed 24 Jan 2024
22. Global Statistical Geospatial Framework, UN Statistics Wiki. https://unstats.un.org/wiki/display/GSGF/Global+Statistical+Geospatial+Framework. Accessed 25 Jan 2024
23. ISO/IEC 19941:2017, Information technology, Cloud computing, Interoperability and portability. https://www.iso.org/standard/66639.html
24. UNECE, Statistics, Generic Statistical Business Process Model (GSBPM) version 5.1. https://unece.org/sites/default/files/2023-11/GSBPM%20v5_1.pdf. Accessed 21 Jan 2024
25. ISO 19170-1:2021, Geographic information—discrete global grid systems specifications—Part 1: core reference system and operations, and equal area earth reference system. https://www.iso.org/standard/32588.html. Accessed 12 Jan 2024
26. ISO/IEC 20547-3:2020, Information technology—big data reference architecture. https://www.iso.org/standard/71277.html. Accessed 9 Oct 2023

Chapter 2
Methods and Concepts About Evaluation

Abstract This chapter discusses the evaluation in the PDCA cycle consisting of plan (P), do (D), check (C), and action (A) for policy making. We will discuss a method that incorporates Bardach's eight steps of policy analysis and performs policy analysis in a data-driven manner. Throughout gap analysis and data analysis to define problems based on evidence obtained at the policy analysis step, we explain how to identify gaps between ideal and current situations. Gap analysis is helpful to compare and analyze policy alternatives. Furthermore, to conduct policy analysis in a data-driven and systematic manner, it is effective to build a data infrastructure for the index values of evaluation criteria. As a data management framework, we introduce the Generalized Statistical Business Process Model (GSBPM), which is practical when an organization produces data and statistics, and KPI management, which is essential when an organization manages its process based on KPIs.

2.1 What is Evidence

The OECD reports the importance of skills and capacities of policy makers for evidence-informed policy-making (EIPM) [1, 2]. To support EIPM, the four prespectives for evidence are addresses:

(1) Necessity of evidence: Evidence is necessary to fight against a post-fact/fake news world, to design effective policies and to assign resources
(2) Meeting demand: we need better understanding for needs of evidence
(3) Evidence mismatch: it is valuable to form trust when building evidence with citizens and users in order to match the supply and demand of evidence
(4) Institutional set-up: the importance to use established ex ant and ex post evaluation processes

Thus, the four perspectives for evidence have led to decision makers' needs, presentation, and priority features for evidence.

Evidence (= objective evidence) may be used as an evaluation criterion after policy implementation, or it may be used for comparative analysis of policy alternatives before policy implementation at the first step of policy proposal. Furthermore, the evidence has a hierarchical structure in terms of reliability.

A.-H. Sato and H. Tsubaki, *Evaluation Platform of Sustainability for Global Systems*,
https://doi.org/10.1007/978-981-97-2296-9_2

For example, five reliability scales of the evidence ("levels of evidence" or "levels of evidence hierarchy") are formed in order from highest to lowest [3, 4]:

- Results of systematic reviews or multiple randomized controlled trials
- Results from experimental studies such as single randomized controlled trials
- Results from observational studies such as before-and-after comparisons
- Results from case studies
- Results of theoretical considerations through the presentation of logic models or expert opinions.

There are several positions on the purpose of policy evaluation. Seeking objective evidence leads to increasing possibility and efficiency of policy implementation by optimizing interim and final outcomes. On the other hand, in actual socioeconomic activities, there is circular causality where cause and effect are almost always causally related to each other. Hence, it is difficult to discriminate causes and effects under noises resulting from other interventions. The generalization of high level of evidence cannot be justified.

Even if the evidence level is low, the level of evidence is raised to the top through natural experiments combined with repeated observational studies of policies [5].

2.2 PDCA Cycle and Data-Driven Evaluation

This section considers the four perspectives of policy planning, implementation, evaluation, and proposal of policy alternatives as the Plan-Do-Check-Action (PDCA) cycle. The PDCA cycle is also known as a Deming cycle, which is a continuous quality improvement model [6–8]. Shewhart proposes a circle of the three steps (specification, production, and inspection) in a dynamic scientific process of acquiring knowledge [6]. Modifying the Shewhart circle, Deming proposes Deming Wheel [7] in the context of business management. Ishikawa redefined the PDCA cycle as a cyclic process consisting of Plan, Do, Check, and Action [8].

When solving regional issues from a sustainability perspective, multiple individuals and organizations must coordinate their activities and interests. There are two approaches to solving regional issues that involve various stakeholders who share interests: (1) methods for solving regional issues through policy guidance and administrative activities carried out by local governments and the national government and (2) methods for solving regional issues that involve stakeholders. There is a method to solve the problem through economic rationality while coordinating with all stakeholders. A policy management cycle has been proposed as a solution through policy guidance and administrative activities [9]. In this cycle, Kaoru Ishikawa's Total Quality Control (TQC) methodology is used in the phase from problem definition to diagnosis (cause analysis) [8] and is considered to be valid.

Stakeholder capitalism, announced at the Davos Conference (World Economic Forum) in January 2020, is a systematized approach based on coordination among stakeholders based on economic rationality. This is a type of capitalism that emphasizes relationships with stakeholders related to a company and emphasizes long-term contributions to stakeholders through corporate activities [10]. The stakeholder capitalism has many common concepts with the Public Interest Capitalism. The Public Interest Capitalism [11, 12] has the following three basic principles such as:

(1) Mid- to Long- term Perspective
(2) Fair Distribution to Contributing Partners (employees, suppliers, customers, the earth, and medium- to long-term shareholders, etc.)
(3) Improvement through entrepreneurial sprits.

In both of these cases, continuous improvement is the best solution for solving regional issues that have arisen due to geopolitical and historical circumstances. In other words, an effective prescription is to implement the PDCA cycle [8] steadily, taught worldwide in 1951 by Kaoru Ishikawa and the first-generation quality control activity creation group in Japan and revised in 1980.

When implementing solutions to regional issues, the following process model as a program science grammar that combines the following problem-solving QC story and the method of continuous improvement using PDCA may function effectively.

(1) Clear definition of desired program performance
(2) Discovering unacceptable violations and exceptions influencing the desired program performance
(3) Logical coverage of symbol strings that may affect its performance
(4) Comparison and evaluation from the viewpoint of necessary and sufficient symbol strings
(5) Implementation of the optimal symbol string into program
(6) Program implementation and monitoring of its performance.

In Japan, it was proposed as a problem-solving QC story in 1960 and was proposed at Komatsu [13] as a standard scenario for improvement activities. The QC Story Methodology consists of 7 steps:

1. The reason for problem choosing
2. Status analysis of the problem
3. Activity Plan
4. Cause-Effect Analysis
5. Problem solution
6. Result confirmation
7. Standardization.

Later, it was standardized as Total Quality Management (TQM) in the Japanese industry [14] and internationally as a process called DMAIC [15] in Six Sigma activities, which has been internationally standardized as ISO 13053-1:2011 [16]. By appropriately placing such continuous improvement activities in the PDCA cycle and implementing them, local issues can be brought closer to resolution, and the ideal form of increasing sustainability can be achieved. It will be possible to manage social issues of sustainability.

When attempting to carry out such activities in a data-driven manner, it is necessary to improve policies using a logic model consisting of inputs, outputs, outcomes, and impacts based on the causal chain of policies. At the same time, policy analysis [17] is required to make proposals possible. This logic model and policy analysis constitute the PDC and CAP of a PDCA cycle and realize continuous improvement through circulation.

Specifically, we emphasize on a method examined in which policy analysis is conducted with sufficient data analysis at the time of policy evaluation. Moreover, policy alternatives (or improvement proposals) are important as policy proposals after the time of evaluation. In this way, we consider a method for continuously improving policy implementation plans in combination with gap analysis.

In particular, the eight-step method in policy analysis proposed by Bardach [17] is positioned as the C-A-P from the PDCA cycle, and it is shown that a cycle can be formed by combining it with the P-D-C of policy planning, implementation, and evaluation based on a logic model.

The evaluation (C: Check) followed by the proposal of an improvement plan (A: Action) is positioned as a "proposal for policy alternatives" in policy analysis. We will discuss the possibility of incorporating data-driven methods into policy evaluation by actively utilizing quantitative and qualitative data analysis in policy analysis.

Furthermore, to make the use of data for policy analysis useful, we propose to develop statistical and data environments (data infrastructure) for the evaluation criteria used in policy evaluation. In addition, evaluation should be "beneficial to the evaluated person." We will also consider how opportunities for policy evaluation can be beneficial to those who are evaluated.

2.3 Policy Evaluation and Logic Model

We explain the logic model that forms the basis of policy evaluation. Figure 2.1 shows the conceptual illustration of a logic model. In the logic model, policies are understood as a chain of inputs, activities, production, outputs, outcomes, and final outcomes (impacts). Input refers to the resources and steps that are used when implementing policy. Activity refers to activities carried out by the government and companies based on input resources, and production refers to changes caused by activities based on input resources. Outputs are the results brought to society by the production, and the outcomes are the changes or benefits that result from the policy implementation. The final outcomes (impacts) are further observed as the changes or benefits in a whole point of view.

Figure 2.1 shows a logic model that expresses a program with inputs as resources, activities to implement the program, outputs of programs, and outcomes and impacts desired in the program. A causes and effects relationship is established from inputs to outcomes and impacts. That corresponds to planning (P), doing (D) and checking (C) in the PDCA cycle. Of course, since the resources to be invested in implementing the policy need to be prepared in advance, a policy plans (inputs) are causes for outputs.

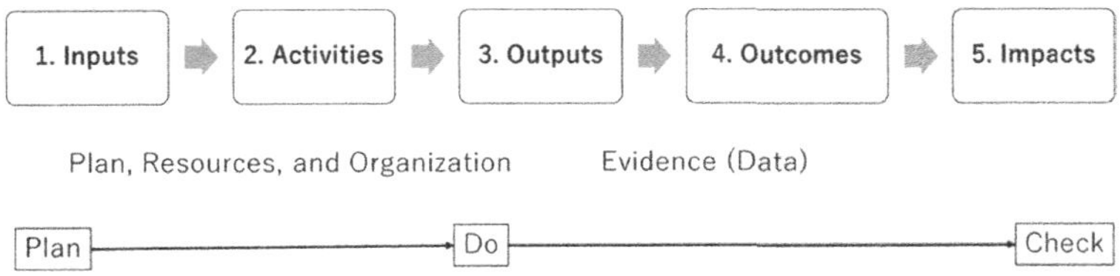

Fig. 2.1 The Logic model that expresses a chain of causes and effects from inputs to outcomes and impacts

They are usually planned together with the budget and the order of activities in advance (P: Plan). The activity is what we do in actual environment (D: Do), and the outputs are observed as changes that occur through activities. We can collect activities and environmental states as evidence (data) and the productions are estimated. The output (intermediate result) or the outcome (impact) is measured as an indicator value of the evaluation criteria that can measure the policy evaluation (C: Check).

2.4 Methods for Policy Analysis

The method of policy analysis is summarized based on the eight steps of policy analysis, which are proposed by Bardach [17]. The eight steps of policy analysis are:

(1) Problem definition
(2) Collection of evidence
(3) Design of policy alternatives
(4) Selection of evaluation criteria
(5) Prediction of results
(6) Comparative analysis of policy alternatives
(7) Decision on policy recommendations
(8) Presentation.

Problem definition is defined as "defining problems that decision-makers can address within their authority and within a limited time, and policy analysis defines problems that can be addressed." As for problem definitions, it is important to note that data analysis of the causal relationship between cause and effect by evidence analysis is effective, and that problems that are directly linked to solutions should not be defined in the problem definition.

It is essential to investigate the history and background of policy, grasp causal relationships, and understand the authority and claims of stakeholders when collecting evidence. The design of policy alternatives is to design logic models consisting of inputs, activities, output, outcomes, and impacts in ways other than the evaluated program. There are four approaches to designing policy alternatives. (1) Present alternatives to existing policies that have been proposed by others. (2) Modify measures adopted in other regions and are considered successful to make them policy alternatives. (3) Propose policy alternatives that correct externalities by tax measures and forced disclosure of information to correct incomplete information markets. (4) It is for the analyst to attempt to form an entirely new alternative program that has never existed before.

Evaluation criteria are criteria for evaluating the expected outcome or process of each policy alternative, and multiple independent or exclusive evaluation criteria should be selected. Evaluation criteria require objectivity, that is, evaluation criteria that lead to the same results no matter who, when, and where they do. Frequently used criteria include efficiency, cost, fairness, political viability, administrative viability, other degrees of freedom, and the safety of life and body.

Outcome prediction predicts the outcome of policy alternatives. Predictions about the future involve uncertainty and caution not to be optimistic, and unexpected side effects need to be considered.

Comparative analysis of policy alternatives includes (a) cost-benefit analysis (cost-benefit analysis converts all evaluations into money), (b) cost-effectiveness analysis (cost-effectiveness analysis selects one evaluation criterion other than money and converts others into money), and (c) multi-attribute analysis (multi-attribute analysis is used when the results of multiple evaluation criteria cannot be converted into money, and the final figures are compared by taking weight according to importance). In general, if a set of independent exclusive evaluation criteria are adopted, it becomes a multi-dimensional optimization problem, so that there is a problem that we face a Pareto optimum to be figured out by various optimization for the evaluation.) One of three types of analysis methods is adopted, and a comparative analysis of policy alternatives is conducted. Some excellent policy alternatives are selected, policy recommendations are determined. Finally, we may make a presentation with an explanation of the process, comparative methods, figures, and validity of the model.

If the eight-step process of policy analysis can be introduced into policy evaluation, it is possible to make policy recommendations and policy alternatives (or improvement proposals) beyond policy evaluation and establish the PDCA cycle at Check and Action with continuous improvement processes.

In the PDCA cycle of policy grogram, there is a part where the plan (P), do (D), check (evaluation) (C) forms a plan and evaluates the results from activities. Moreover, there is a part where we analyze the current situation and improve the current plan by recognizing gaps and recommendation of policy alternatives. This may form check (evaluation) (C), action (improvement) (A), and plan (P).

Let us consider relationships between the logic model shown in Fig. 2.1 and the PDCA cycle shown in Fig. 2.2. The logic model in Fig. 2.1 corresponds to the parts of planning (P), doing (D), checking (evaluation) (C). As shown in Fig. 2.2, a logic model is placed as PDC in the PDCA cycle, the parts of the checking (evaluation) (C), action (improvement) (A), and plan (P) should be supplemented as shown in Fig. 2.3. The PDCA cycle is described as a circular flow that planning (P), doing (D), checking (C), action (improvement) (A), and plan (P) are repeated.

In particular, for improving the evaluated policy plan, it is important that data analysis is used for recognizing the gap between the current situation and the ideal.

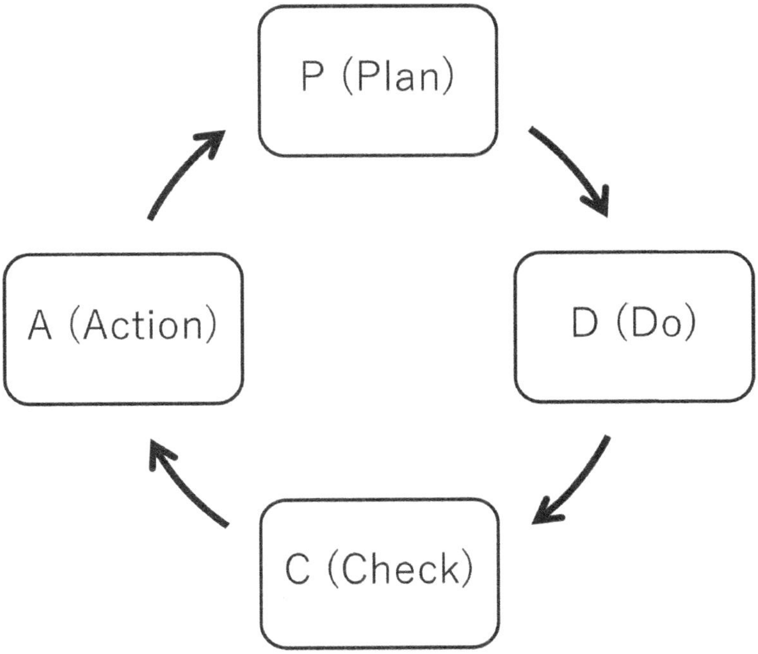

Fig. 2.2 A graphical representation of the PDCA cycle, which consists of planning (P), doing (D), checking (C), and acting (A)

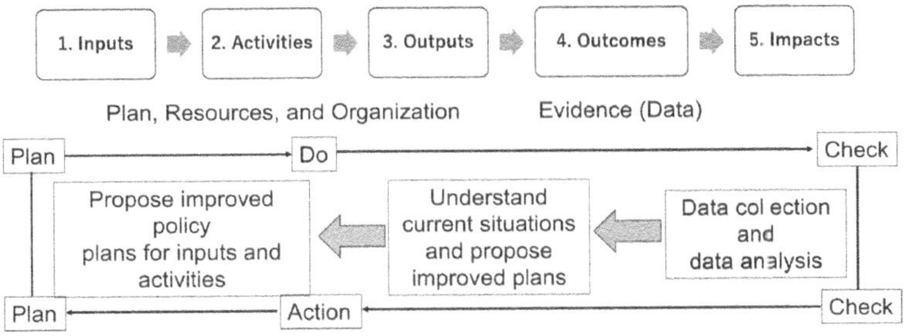

Fig. 2.3 A modified logic model composed by combining a logic model and eight steps of the policy analysis. The original logic model corresponds to P-D-C of the PDCA cycle. The eight steps of the policy analysis forms C-A-P of the PDCA cycle for the logic model. The circular form of the logic model may improve the outcomes and impacts of the program eventually if we adjust the gap between current outcomes (impacts) and our future vision adequately

Furthermore, by recognizing the gap (the difference between the current and ideal situations), it is necessary to carry out improvement activities to change behavior so that the gap decreases.

2.5 Data Analysis Based on Gap Analysis

2.5.1 *Gap Analysis*

Gap analysis is a method of identifying items necessary to eliminate the gap between ideal (To Be) and reality (As Is). By recognizing the gap between the ideal and the reality and repeating the question of why such a gap is occurring, it is possible to approach the cause and effect which are extracted at the problem definition.

The plan is improved by changing inputs and activities that can solve the problem extracted at the gap analysis. We may expect to observe that some kind of gap reduction occurs in the next evaluation by implementing it.

Data analysis based on gap analysis is performed by the following steps:

1. Think about expected status (To Be)
2. Clarify the current situation (As Is) from data analysis
3. Identify issues by recognizing gaps between To Be and As Is and searching for causes
4. Determine actionable solutions to challenges
5. Incorporate actionable solutions into plan changes.

The gap analysis is important step to have an ideal state in future and to clarify the current situation from data analysis. Therefore, it is desirable to actively utilize data and quantitatively grasp gaps for gap analysis.

2.5.2 *Data Analysis for Gap Analysis*

In gap analysis, data analysis is extremely effective because it is necessary to quantitatively understand the gap between the ideal vision and the current situation. This corresponds to problem definition and evidence collection in the eight-steps of policy analysis. Data analysis consists of the following five steps:

1. Data acquisition: It is a task to acquire necessary data for analysis and not to attempt to get all the datasets at once. The point of data acquisition is to acquire a small amount of the necessary data at first.
2. Data understanding: It is a task to deepen understanding of data and the context where data is produced.
3. Data preparation: It is the task to prepare data into a form that can be visualized and analyzed by computer software.

4. Data visualization and analysis: It is the task to draw diagrams and compute indicators to represent characteristics of phenomena captured by the data from prepared data. The aim of data visualization and analysis is to understand and explain insights obtained through data analysis immediately.
5. Data interpretation: It is the task of telling the meaning from the obtained diagrams and indicators. Stories are important for understanding and explaining meaning.

Therefore, in order to enable policy analysis to be data-driven and efficient, it is necessary to construct data infrastructure of data and statistics and the continuous collection of data to compute indicators to be used as evaluation criteria. Some references show that 80% of data analysis efforts are consumed for pre-processing such as data acquisition, data understanding, and data processing [18]. Moreover, some indicators that have not been disseminated yet are newly developed and attempted to be designed, implemented, and disseminated. Thus, it is important to construct a data platform to collect, store, analyze, dessiminate data.

2.6 Management Framework

2.6.1 *Generic Statistical Business Process Model*

Generic Statistical Business Process Model (GSBPM) [19] is a useful framework that is proposed by the United Nations European Commission (UNECE) and its documents are being prepared sequentially. As shown in Fig. 2.4, GSBPM is a flow of activities used at the time when we produce data and statistics and decomposed into eight phases. Each phase is further decomposed into multiple steps as mentioned in Chap. 1.

GSBPM includes the process consisting of Specify Needs, Design, Build, Collect, Process, Analyze, Disseminate, Archive, and Evaluate in a social context. It is divided into three phases: research and development phase, implementation and deployment phase, and social effects and impacts. Furthermore, the evaluation results are returned to the research and development phase and used for improvement.

In practice, it is difficult for a single person to complete the work contained in the nine phases required by the generalized business process model. Therefore, it should be carried out across multiple people or organizations.

Figure 2.5 shows the GSBPM with roles for eight phases. There are four types of roles: data and statistics providers, data and statistics users, researchers and developers, and statistics and data conductors. Researchers and developers are mainly involved in the setting, design, and specify of needs during the research and development phase. Data and statistics providers are responsible for specifying needs, designing, building, collecting, processing, analyzing, disseminating, archiving, and evaluating. Data and statistics users are responsible for collecting, processing, analyzing, disseminating, and evaluating. Statistics and data orchestrators are involved in the entire system and are involved in the ongoing coordination of the work of

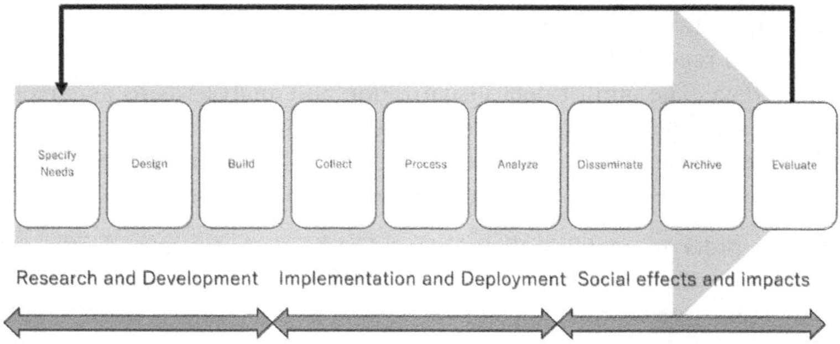

Fig. 2.4 A sequence of nine phases illustrates the generic statistical business process model, which consists of specify needs, design, build, collect, process, analyze, disseminate, archive, and evaluate

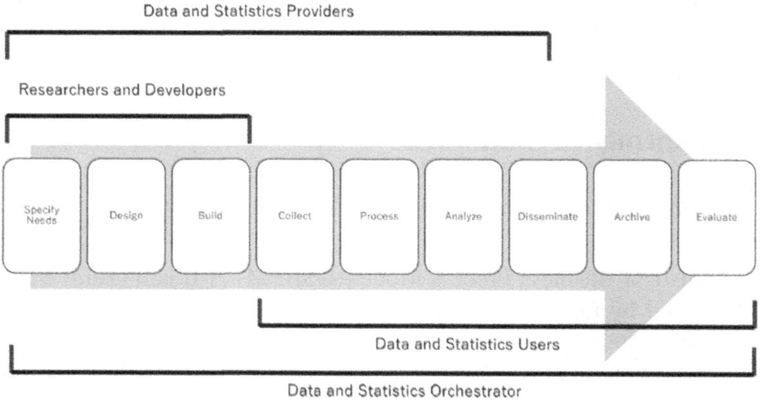

Fig. 2.5 The generic statistical business process model with four roles of researchers and developers, data and statistics users, data and statistics providers, and data and statistics orchestorators

researchers, developers, statistics and data providers, and statistics and data users by coordinating with institutional design and organizational management.

In recent years, the development of ICT promoting users' participation in the collection, processing, analysis, and publication. The ICT's technical advantage may create an overlap between providers of statistics and data and users of statistics and data. In other words, it can be regarded that there have been technological developments that allow people who are users of statistics and data as policymakers to be greatly involved in the collection, processing, and analysis of statistics and data. If there is an autonomous decentralized evaluation platform, policy implementation can be efficient, and it may increase a possibility for improvement activities to enhance sustainability.

2.6.2 *KPI Management*

Moreover, the Key Performance Indicator (KPI) monitors activities to manage operations quantitatively. It is a typical method of managing organizations and is called KPI management [20]. The KPI management consists of four essential concepts: Goal, Key Goal Indicator (KGI), Key Success Factor (KSF), and KPI. Figure 2.6 shows a relationship among Goal, KGI, KSF, and KPI in the KPI management.

The Goal is the purpose of an activity expressed as a sentence. In general, commercial organizations often determine that the Goal is an optimization of an amount of sales or profit (sales—costs). In this case, KGI is set as sales or profit. Furthermore, sales and profits are key components in organization's financial report information. Thus, such KGIs is specifically called as the Key Financial Indicator (KFI). KSF should be the most critical process of an organization's business or operational processes. Thus, if organizations determine KSF as the weakest process in their own process, they strengthen their process by improving actions of the weakest process in their behavior from an organizational point of view. The KPI is a quantitative measure to express actions. If we can set KPI as an indicator that describes KSF adequately, the correlation between KGI and KPI becomes high, and KPI proceeds from KGI.

Thus, increasing (or decreasing) KPI is equivalent to achieving our Goal. The critical point of the KPI is a strong correlation with the KGI and one of the indicators leading to the KGI. If these two properties are satisfied, a process management based on the KPI leads to achieving the Goal.

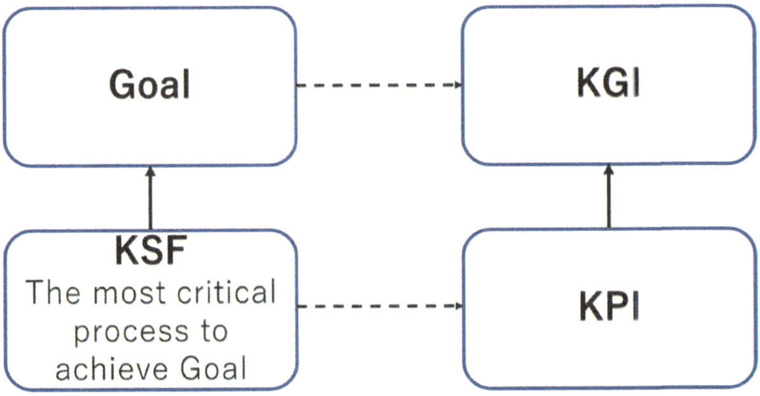

Fig. 2.6 The relationship among Goal, KGI, KSF, and KPI in KPI management

2.7 Conclusion

This chapter considered levels of evidence and mentioned the relationship between a logic model and PDCA cycle. The logic model consists of inputs, activities, outputs, outcomes, and impacts. Policy makers conducts policy implementation using plan, resources, and organization. During implementation, evidence and data are produced. The outputs and outcomes are measured by indices to express social contexts. The final outcomes are obtained from outcomes produced by several outputs. The logic model corresponds to P-D-C in the PDCA cycle. Collecting and analyzing data accumulated through policy implementation improve our understanding of current situations and enable us to propose improved plans. We can use insights obtained by data analysis for Action (improvement) based on the eight steps of policy analysis. Data infrastructure for policy analysis is useful for collection of evidence and selection of evaluation criteria. This part corresponds to C-A-P in the PDCA cycle. Therefore, the PDCA cycle for policy making can be realized by combining the logic model and policy analysis. We call such activities data-driven decision-making.

In this case, GSBPM consists of nine phases (Specify Needs, Design, Build, Collect, Process, Analyze, Disseminate, Archive, and Evaluate) and enables different people and organizations to share and cooperate processes to produce and use data and statistics. We propose four types of roles: researchers and developers, data and statistics providers, data and statistics users, and data orchestrators. We discuss how to collaborate on the four roles in GSBPM. Furthermore, KPI management is practical for monitoring activities and achieving goals, which helps manage organizations in a data-driven manner.

In recent years, the development of ICT promoting users' participation in the collection, processing, analysis, and publication. The ICT's technical advantage may create an overlap between data and statistics provides and data and statistics users. In other words, it can be regarded that there have been technological developments that allow people who are data and statistics users as policy makers to be greatly involved in the collection, processing, and analysis of statistics and data.

If there is an autonomous decentralized evaluation platform, policy implementation can be efficient, and it may increase a possibility for improvement activities to enhance sustainability.

References

1. OECD, *Building Capacity for Evidence-Informed Policy-Making: Lessons from Country Experiences*. https://doi.org/10.1787/86331250-en. Accessed 28 Dec 2023
2. OECD Conference Centre, Governing better through evidence-informed policy making, Conference summary, 26–27 June 2017. https://www.oecd.org/gov/governing-better-through-evidence-informed-policy-making-proceedings.pdf
3. B. Patricia, R. Burns, J. Rohrich, K.C. Chung, The levels of evidence and their role in evidence-based medicine. Plast. Reconstr. Surg. **128**(1), 305–310 (2011)

4. Levels of evidence in research. https://scientific-publishing.webshop.elsevier.com/research-process/levels-of-evidence-in-research/. Accessed 31 Dec 2023
5. P. Craig, C. Cooper, D. Gunnell, S. Haw, K. Lawson, S. Macintyre, D. Ogilvie, M. Petticrew, B. Reeves, M. Sutton, S. Thompson, Using natural experiments to evaluate population health interventions: new medical research council guidance. J. Epidemiol. Community Health **66**(12), 1182–1186 (2012)
6. W.A. Shewhart, *Statistical Method from the Viewpoint of Quality Control*. The Graduate School, The Department of Agriculture, Washington (1939)
7. W.E. Deming, *Elementary Principles of the Statistical Control of Quality* (JUSE, 1950)
8. K. Ishikawa, *What is Total Quality Control? The Japanese Way*, Translated by D.J. Lu (Prentice-Hall, Inc., Englewood Cliffs, NJ, 1985), pp. 56–61
9. M.R.W. Hsiao, P. Berman, M. Reich, *Getting Health Reform Right: A Guide to Improving Performance and Equity* (Oxford University Press, 2008)
10. World Economic Forum, THE DAVOS AGENDA 2021, What is stakeholder capitalism? https://www.weforum.org/agenda/2021/01/klaus-schwab-on-what-is-stakeholder-capitalism-history-relevance/. Accessed 28 Oct 2023
11. G. Hara, Public Interest Capitalism, Bunshun Shinsho, Tokyo (2000). (in Japanese)
12. The Alliance Forum Foundation, Public Interest Capitalism. http://www.allianceforum.org/en/capitalism/. Accessed 28 Dec 2023
13. Komatsu Co Ltd., https://www.komatsu.jp/en, Accessed 26 Dec 2023
14. R. Moen, C. Norman, Evolution of the PDCA cycle, in *Proceedings of the 7th ANQ Congress, 17 Sept 2009*. https://elfhs.ssru.ac.th/phusit_ph/pluginfile.php/48/block_html/content/Moen-Norman-2009%20PDCA.pdf. Accessed 28 Oct 2023
15. AQC, DMAIC, https://asq.org/quality-resources/dmaic. Accessed 7 Jan 2024
16. ISO 13053-1:2011, Quantitative methods in process improvement—Six Sigma—Part 1: DMAIC methodology
17. E. Bardach, *A Practical Guide for Policy Analysis: The Eightfold Path to More Effective problem Solving*, 3rd edn. (CQ Press, 2009)
18. A. Ruiz, The 80/20 data science dilemma, InfoWorld (2017). https://www.infoworld.com/article/3228245/the-80-20-data-science-dilemma.html. Accessed 21 Jan 2024
19. UNECE, Statistics, Generic Statistical Business Process Model (GSBPM) version 5.1. https://unece.org/sites/default/files/2023-11/GSBPM/%20v5_1.pdf. Accessed 21 Jan 2024
20. D. Parmenter, *Key Performance Indicators: Developing, Implementing, and Using Winning KPIs*, 4th edn. (Wiley, Hoboken, NJ, 2020)

Chapter 3
Grid Squares and Grid Square Statistics

Abstract This chapter introduces some examples of Grid Square and Grid Square statistics concepts. The purpose of this chapter is to introduce examples of Grid Square statistics and to help readers deepen their understanding of Grid Square statistics. Grid Square statistics need a standardized Grid Square coding system. This chapter introduces several definitions of Grid Square coding system and compares their strength and weakness. Moreover, we explain the Japanese National Grid Square codes (JIS X0410:2002) standardized in the Japan Industrial Standards (JIS). The Japanese National Grid Square statistics have been developed since the 1960s for both industrial and government usage in Japan. Finally, we show the definition of the World Grid Squares as an extension of JIS X0410:2002 for worldwide usage.

3.1 Grid Squares

Grid Square statistics was developed as one of the analysis methods in geography, producing statistics using tiny rectangular areas as aggregation units.

Pure Geography [1] written by Finnish geographer Granö tells us that geography is to accurately express, understand, and interpret the human environment. In particular, it discusses geography as a method for spatially expressing the environment based on human perception, interpreting regional characteristics, distinguishing areas, and classifying them according to their proximity.

Spatial statistics are statistics aggregating quantitative or qualitative data linked to geospatial units. In the case of Grid Square statistics, statistics are produced in large quantities using tiny rectangular areas (called Grids or Grid Squares) as aggregation units in space. This is a remarkable characteristic of Grid Square statistics.

Generally, spatial statistics are produced using administrative divisions (province, state, prefecture, city, and municipality) as aggregation units, but administrative divisions may change over time. Thus, statistics and data regarding administrative divisions can only be compiled over a short period. This becomes a critical issue when comparing or reaggregating spatial statistics regarding administrative divisions.

Grid Square statistics are determined from location information, such as latitude and longitude, which changes relatively little in time. Therefore, we can compare

A.-H. Sato and H. Tsubaki, *Evaluation Platform of Sustainability for Global Systems*,
https://doi.org/10.1007/978-981-97-2296-9_3

statistics and data over a long period. Furthermore, it is possible to reconstruct arbitrary aggregation units by re-aggregation. It has the characteristics of Grid Square statistics better than spatial statistics regarding administrative divisions.

The United Kingdom is said to be the first country to use the Grid Square method for official statistics, and it is stated that Grid Square statistics have been produced under the Modified British Grid System since 1931.

The benefits of Grid Square statistics are realized when it is standardized and used by many people, organizations, and companies, such as industrial or national standards. Establishing Grid Square statistics requires a comprehensive system of systems consisting of laws, mathematics, and computational methods. As a result, only a few countries in the world have been able to establish Grid standards as part of their national standards, including the United Kingdom [2], Japan [3], the European Union [4], South Korea [5], and Australia [6].

3.2 Existing Grid Coordinate Systems

Currently, there are several types of Grid Squares have been already defined in some countries and organizations.

The first Grid Square coding system was established in the United Kingdom (UK) in 1938 [2]. It is currently called the British National Grid (BNG) or the Ordnance Survey National Grid and has been widely used in the UK to generate Grid Square statistics from national statistical surveys.

These statistics are calculated from latitude and longitude based on the Universal Transverse Mercator (UTM) coordinate system, which sets 49°N and 2°W as the true origin. The BNG, therefore, has an available range to be used as a grid reference.

The vertical lines are called eastings, and they increase in value from west to east. The horizontal lines are called northings as they increase in values from south to north.

The unique reference system can be applied to all BNG maps of Great Britain, at all scales. Figure 3.1 shows a conceptual illustration of the Ordnance Survey National Grid.

Great Britain is covered by Grid Squares measuring 100 km across. Each Grid Square is identified by two letters. One 100 km Grid Square consists of 100 10 km grids. A set of eastings numbers (0–9) and northtings numbers (0–9) indicate a 10 km grid. Moreover, one 10 km grid square consists of 100 1 km grids. A set of eastings (0–9) and northtings (0–9) numbers indicate a 1 km grid. Therefore, 1 km grid square is identified as two letters plus four decimal digits.

More recently, the European Union (EU) Commission has proposed European Reference Grids [4] and has used them to generate Grid Square statistics about population census in EU member states [7]. This is known as GEOSTAT 1A and provides a census population prototype dataset after 2006.

The European Reference Grids are calculated from latitude and longitude based on the ETRS89 Lambert Azimuthal Equal-Area Projection Coordinate Reference Sys-

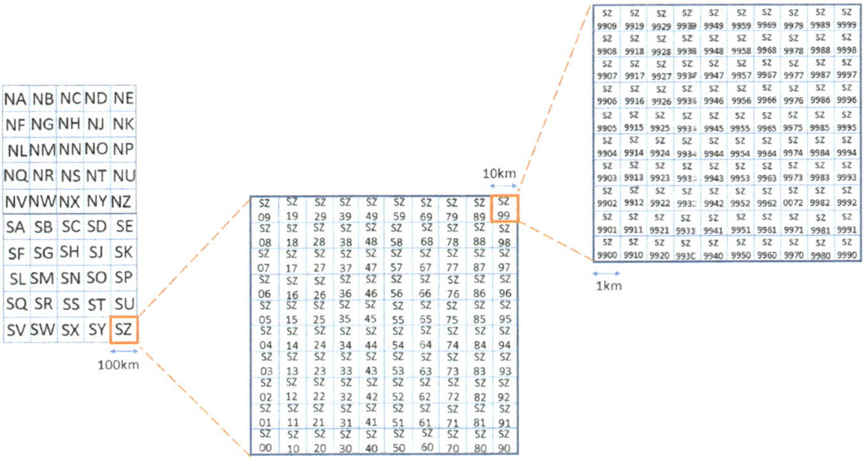

Fig. 3.1 A conceptual illustration of ordnance survey national grid

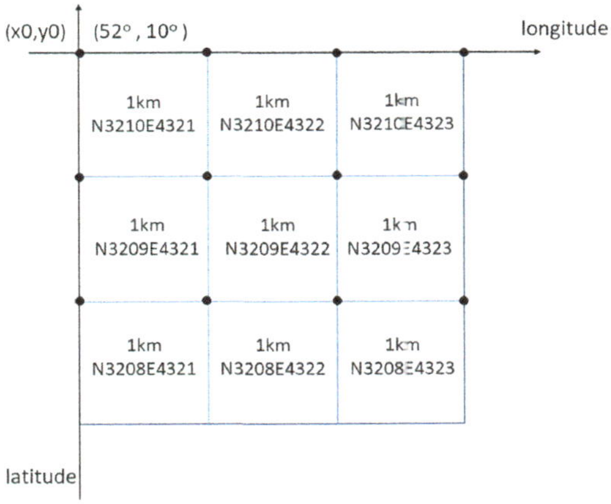

Fig. 3.2 A conceptual illustration of the European reference grids

tem (ETRS89/ETRS-LAEA), which uses a projection point set at 52°N and 10°E. There is a hierarchical structure of Grid resolutions. For example, the size of Grid Square is defined for 1, 10, 100, 1,000, 10,000 and 100,000 m. Figure 3.2 shows an example of 1 km Grid Square of the European Reference Grids around the projection point 52°N and 10°E. The origin is set as $x_0 = 4321,000$m and $y_0 = 3210,000$m. 1 km Grid Square at the corner closest to the projection point is displayed as 1km N 3210E4321. The European grid reference system is designed based on the Recommendation EUR Report 21494 En:39-46.

The GEOSTAT 1B [8] was produced based on a selected structure as proposed by the Generic Statistical Business Process Model (GSBPM) [10].

Table 3.1 Map projections for Mainland Australia and Tasmania, New Zealand defined in the national nested grids by ANZLIC.

Region	Projection title	Projection EPSG code	Point of origin
Mainland Australia and Tasmania	Australian equal area albers	3577	−2690000, −1000000
New Zealand	New Zealand Transverse Mercator 2000	2193	9830000, 6217000
Antarctica	Antarctic polar stereographic	3031	−3000000, −3000000

The Australian Bureau of Statistics produces Grid Square statistics of census population [9] based on the National Nested Grid by the Australia New Zealand Land Information Council (ANZLIC) [6]. The The ANZLIC's National Nested Grid (NNG) defines Grid Squares (cells) in Mainland Australia and Tasmania, New Zealand Mainland, and Antarctica as shown in Table 3.1.

Since the BNG, European Reference Grids, and ANZLIC National Nested Grids are based on the Equal-Area Projection Coordinate Reference System, they cannot be extended for worldwide usage due to their applicable projected bounds.

3.3 Japanese National Grid Square Coding System

A geographic standard for Official Statistics in Japan as Grid Squares date back to the middle of 1960s. In 1973, "Standard Grid Square and Grid Square Code Used for the Statistics" was made as the Announcement No. 143 of the Administrative Management Agency (AMA) that hierarchically defines Grid Squares covering the entire land of Japan. It has been adopted as Japanese Industrial Standards (JIS) JIS C 6304:1976 (currently X0410:2002) since 1976.

The Japanese National Grid Square codes (JIS X0410:2002) are computed from latitude and longitude based on the following equations:

$$\text{1st–level grid square code} = pu \tag{3.1}$$

$$\text{2nd–level grid square code} = puqv \tag{3.2}$$

$$\text{3rd–level grid square code} = puqvrw \tag{3.3}$$

$$\text{1/2 grid square code} = puqvrws_2 \tag{3.4}$$

$$\text{1/4 grid square code} = puqvrws_2s_4 \tag{3.5}$$

$$\text{1/8 grid square code} = puqvrws_2s_4s_8 \tag{3.6}$$

where integers p, u, q, v, r, w, s_2, s_4, and s_8 are computed from the following equations.

$$p := \lfloor \text{latitude} \times 60/40 \rfloor \tag{3.7}$$
$$a := (\text{latitude} \times 60/40 - p) \times 40 \tag{3.8}$$
$$q := \lfloor a/5 \rfloor \tag{3.9}$$
$$b := (a/5 - q) \times 5 \tag{3.10}$$
$$r := \lfloor b \times 60/30 \rfloor \tag{3.11}$$
$$c := (b \times 60/30 - r) \times 30 \tag{3.12}$$
$$s_{2u} := \lfloor c/15 \rfloor \tag{3.13}$$
$$d := (c/15 - s_{2u}) \times 15 \tag{3.14}$$
$$s_{4u} := \lfloor d/7.5 \rfloor \tag{3.15}$$
$$e := (d/7.5 - s_{4u}) \times 7.5 \tag{3.16}$$
$$s_{8u} := \lfloor e/3.75 \rfloor \tag{3.17}$$
$$u := \lfloor \text{longitude} - 100 \rfloor \tag{3.18}$$
$$f := \text{longitude} - 100 - u \tag{3.19}$$
$$v := \lfloor f \times 60/7.5 \rfloor \tag{3.20}$$
$$g := (f \times 60/7.5 - v) \times 7.5 \tag{3.21}$$
$$w := \lfloor g \times 60/45 \rfloor \tag{3.22}$$
$$h := (g \times 60/45 - w) \times 45 \tag{3.23}$$
$$s_{2l} := \lfloor h/22.5 \rfloor \tag{3.24}$$
$$i := (h/22.5 - s_{2l}) \times 22.5 \tag{3.25}$$
$$s_{4l} := \lfloor i/11.25 \rfloor \tag{3.26}$$
$$j := (i/11.25 - s_{4l}) \times 11.25 \tag{3.27}$$
$$s_{8l} := \lfloor j/5.625 \rfloor \tag{3.28}$$
$$s_2 := 2s_{2u} + s_{2l} + 1 \tag{3.29}$$
$$s_4 := 2s_{4u} + s_{4l} + 1 \tag{3.30}$$
$$s_8 := 2s_{8u} + s_{8l} + 1, \tag{3.31}$$

where $\lfloor x \rfloor$ is the floor function, which is defined as the maximum integer less than or equal to x. For example, $\lfloor 1.4 \rfloor = 1$, $\lfloor 5.7 \rfloor = 5$, and so on.

On the contrary, if we have each level of Grid Square code as shown in Fig. 3.3, geographical location (latitude and longitude) of four corners of the given Grid Square such as

northern-east corner: (latitude_n, $\text{longitude}_\text{w}$)
southern-west corner: (latitude_n, $\text{longitude}_\text{w}$)
southern-east corner: (latitude_n, $\text{longitude}_\text{w}$)
northern-west corner: (latitude_n, $\text{longitude}_\text{w}$)

can be identified by using the four values: latitude_n, $\text{longitude}_\text{w}$, latitude_s, $\text{longitude}_\text{e}$. The geographical location of each level of Grid Square code is computed as follows.

The 1st-level of Grid Square code:

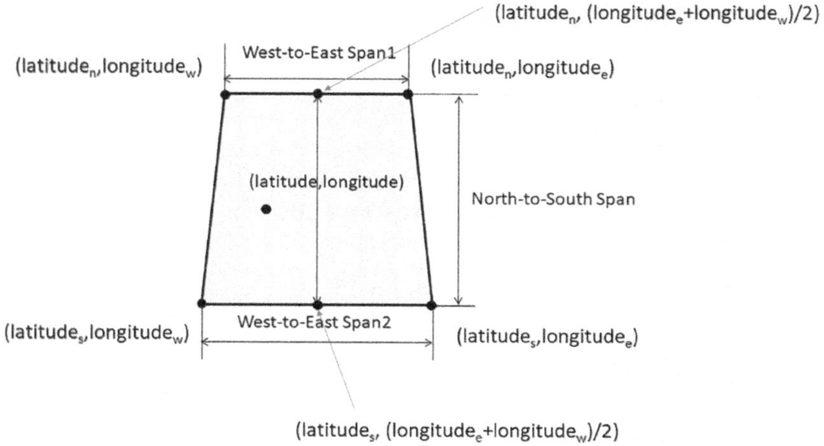

Fig. 3.3 The definition of grid square

$$\text{latitude}_s = p \times 40/60 \tag{3.32}$$
$$\text{longitude}_w = 100 + u \tag{3.33}$$
$$\text{latitude}_n = (p - 1) \times 40/60 \tag{3.34}$$
$$\text{longitude}_e = 100 + u + 1 \tag{3.35}$$

The 2nd-level of Grid Square code:

$$\text{latitude}_s = p \times 40/60 + q \times 5/60 \tag{3.36}$$
$$\text{longitude}_w = 100 + u + v \times 7.5/60 \tag{3.37}$$
$$\text{latitude}_n = p \times 40/60 + (q - 1) \times 5/60 \tag{3.38}$$
$$\text{longitude}_e = 100 + u + (v + 1) \times 7.5/60 \tag{3.39}$$

The 3rd-level of Grid Square code:

$$\text{latitude}_s = p \times 40/60 + q \times 5/60 + r \times 30/3600 \tag{3.40}$$
$$\text{longitude}_w = 100 + u + v \times 7.5/60 + w \times 45/3600 \tag{3.41}$$
$$\text{latitude}_n = p \times 40/60 + q \times 5/60 + (r - 1) \times 30/3600 \tag{3.42}$$
$$\text{longitude}_e = 100 + u + v \times 7.5/60 + (w + 1) \times 45/3600 \tag{3.43}$$

The 1/2 Grid Square code:

$$\begin{aligned}\text{latitude}_s &= p \times 40/60 \\ &\quad + q \times 5/60 + r \times 30/3600 + (s_2 - 2\lfloor s_2/2 \rfloor) \times 15/3600\end{aligned} \tag{3.44}$$

$$\begin{aligned}\text{longitude}_w &= 100 + u + v \times 7.5/60 \\ &\quad + w \times 45/3600 + \lfloor s_2/2 \rfloor \times 22.5/3600\end{aligned} \tag{3.45}$$

$$\begin{aligned}\text{latitude}_n &= p \times 40/60 + q \times 5/60 + r \times 30/3600 \\ &\quad + (s_2 - 2\lfloor s_2/2 \rfloor - 1) \times 15/3600\end{aligned} \tag{3.46}$$

$$\begin{aligned}\text{longitude}_e &= 100 + u + v \times 7.5/60 \\ &\quad + w \times 45/3600 + (\lfloor s_2/2 \rfloor + 1) \times 22.5/3600\end{aligned} \tag{3.47}$$

The 1/4 Grid Square code:

$$\begin{aligned}\text{latitude}_s &= p \times 40/60 + q \times 5/60 \\ &\quad + r \times 30/3600 + ((s_2 - 1) - 2\lfloor (s_2 - 1)/2 \rfloor) \\ &\quad \times 15/3600 + (s_4 - 2\lfloor s_4/2 \rfloor) \times 7.5/3600\end{aligned} \tag{3.48}$$

$$\begin{aligned}\text{longitude}_w &= 100 + u + v \times 7.5/60 \\ &\quad + w \times 45/3600 + \lfloor (s_2 - 1)/2 \rfloor \times 22.5/3600 \\ &\quad + \lfloor s_4/2 \rfloor \times 11.25/3600\end{aligned} \tag{3.49}$$

$$\begin{aligned}\text{latitude}_n &= p \times 40/60 + q \times 5/60 \\ &\quad + r \times 30/3600 + ((s_2 - 1) - 2\lfloor (s_2 - 1)/2 \rfloor - 1) \times 15/3600 \\ &\quad + (s_4 - 2\lfloor s_4/2 \rfloor - 1) \times 7.5/3600\end{aligned} \tag{3.50}$$

$$\begin{aligned}\text{longitude}_e &= 100 + u + v \times 7.5/60 \\ &\quad + w \times 45/3600 + \lfloor (s_2 - 1)/2 \rfloor \times 22.5/3600 \\ &\quad + (\lfloor s_4/2 \rfloor + 1) \times 11.25/3600\end{aligned} \tag{3.51}$$

The 1/8 Grid Square code:

$$\begin{aligned}\text{latitude}_s &= p \times 40/60 + q \times 5/60 \\ &\quad + r \times 30 \div 3600 + ((s_2 - 1) - 2\lfloor (s_2 - 1)/2 \rfloor) \times 15/3600 \\ &\quad + ((s_4 - 1) - 2\lfloor (s_4 - 1)/2 \rfloor) \times 7.5/3600 \\ &\quad + (s_8 - 2\lfloor s_8/2 \rfloor) \times 3.75/3600\end{aligned} \tag{3.52}$$

$$\begin{aligned}\text{longitude}_w &= 100 + u + v \times 7.5/60 + w \times 45/3600 \\ &\quad + \lfloor (s_2 - 1)/2 \rfloor \times 22.5/3600 + \lfloor (s_4 - 1)/2 \rfloor \times 11.25/3600 \\ &\quad + \lfloor s_8/2 \rfloor \times 5.625/3600\end{aligned} \tag{3.53}$$

$$\text{latitude}_n = p \times 40/60 + q \times 5/60$$

$$+ r \times 30/3600 + ((s_2 - 1) - 2\lfloor (s_2 - 1)/2 \rfloor) \times 15/3600$$
$$+ ((s_4 - 1)/2 - 2\lfloor (s_4 - 1)/2 \rfloor) \times 7.5/3600$$
$$+ (s_8 - 2\lfloor s_8/2 \rfloor) \times 3.75/3600 \quad (3.54)$$

$$\text{longitude}_e = 100 + u + v \times 7.5/60 + w \times 45/3600$$
$$+ \lfloor (s_2 - 1)/2 \rfloor \times 22.5/3600 + \lfloor (s_4 - 1)/2 \rfloor \times 11.25/3600$$
$$+ (\lfloor s_8/2 \rfloor + 1) \times 5.625/3600 \quad (3.55)$$

The JIS X0410:2002 has hierarchical structure consisting of 6 levels. Administrative areas are used as geographical unit in statistical aggregation, however, they vary over time. Grid square statistics is a geographical statistics defined by latitude and longitude.

Moreover, the Japanese National Grid Square coding system can be extended for worldwide usage due to its simple projection. The World Grid Squares coding system is an extension having compatibility with JIS X0410:2002. This is a key definition to produce Grid Square statistics data from a global perspective.

An alternative World Grid Square coding system is OGC Discrete Global Grid System (DGGS) Standard has been developed based on ellipsoidal polygons constituting the boundary of a DGGS cell, which posses hierarchical structure by dividing a parent cell into several child cells [11]. This procedure has been widely studied in geoscience and a similar concept is introduced into the H3 as known as the Uber's Hexagonal Hierarchical Spatial Index [12].

3.4 World Grid Square Codes as an Extension of Japanese National Grid Square Codes

Recently, three of the authors has proposed the World Grid Square codes in [14]. The World Grid Square coding system adds two upper numeric digits to Japanese National Grid Square codes (JIS X0410:2002). In this section, we show the definition to calculate Grid Square codes from given geographic information (latitude and longitude) and equations to compute four representative positions (northern-east, southern-west, southern-east, and northern-west corner points) of each Grid Square.

3.4.1 Definition of Grid Square Codes

We define the 0th-level Grid Square code as

$$o = 2^2 x + 2y + z + 1, \quad (3.56)$$

where three binary variables x, y, and z separate the earth into eight areas based on latitude and longitude, thus constructing the 0th-level World Grid Square code. The three binary variables x, y, and z are given as follows:

1. Variable $x = 0$ if latitude is positive, otherwise $x = 1$
2. Variable $y = 0$ if longitude is positive, otherwise $y = 1$
3. Variable $z = 0$ for $|\text{longitude}| < 100°$, otherwise $z = 1$.

From 0th-level World Grid Square code o, we obtain x, y, and z, as

$$z = (o - 1) \bmod 2, \tag{3.57}$$

$$y = ((o - z - 1) \div 2) \bmod 2, \tag{3.58}$$

$$x = (o - 2y - z - 1)/4. \tag{3.59}$$

From the geographic information described as latitude and longitude, we can calculate World Grid Square codes, including the positions for six levels. Assume integers p (p is two or three digits), q (q is one digit), r (r is one digit), u (u is one or two digits), v (v is one digit), w (w is one digit), s_2 (s_2 is one digit), s_4 (s_4 is one digit), and s_8 (s_8 is one digit), which are calculated from latitude and longitude as follows:

$$\begin{cases}
p := \lfloor (1 - 2x)\text{latitude} \times 60/40 \rfloor, \\
a := \left\{ (1 - 2x)\text{latitude} \times 60/40 - p \right\} \times 40, \\
q := \lfloor a/5 \rfloor, \\
b := (a/5 - q) \times 5, \\
r := \lfloor b \times 60/30 \rfloor, \\
c := (b \times 60/30 - r) \times 30, \\
s_{2u} := \lfloor c/15 \rfloor \quad (s_{2u} \text{ is one digit}), \\
d := (s_{2u}/15 - s_{2u}) \times 15, \\
s_{4u} := \lfloor d/7.5 \rfloor \quad (s_{4u} \text{ is one digit}), \\
e := (d/7.5 - s_{4u}) \times 7.5, \\
s_{8u} := \lfloor e/3.75 \rfloor \quad (s_{8u} \text{ is one digit}), \\
u := \lfloor (1 - 2y)\text{longitude} - 100z \rfloor, \\
f := (1 - 2y)\text{longitude} - 100z - u, \\
v := \lfloor f \times 60/7.5 \rfloor, \\
g := (f \times 60/7.5 - v) \times 7.5, \\
w := \lfloor g \times 60/45 \rfloor, \\
h := (g \times 60/45 - w) \times 45, \\
s_{2l} := \lfloor h/22.5 \rfloor \quad (s_{2l} \text{ is one digit}), \\
i := (s_{2l}/22.5 - s_{2l}) \times 22.5, \\
s_{4l} := \lfloor i/11.25 \rfloor \quad (s_{4l} \text{ is one digit}), \\
j := (s_{4l}/11.25 - s_{4l}) \times 11.25, \\
s_{8l} := \lfloor j/5.625 \rfloor \quad (s_{8l} \text{ is one digit}), \\
s_2 := s_{2u} \times 2 + s_{2l} + 1, \\
s_4 := s_{4u} \times 2 + s_{4l} + 1, \\
s_8 := s_{8u} \times 2 + s_{8l} + 1
\end{cases} . \tag{3.60}$$

Thus, the 1st-level World Grid Square code (six numeric digits) can be computed as

$$\text{1st-level World Grid Square code} = \begin{cases} o00p0u & (p < 10, u < 10) \\ o0p0u & (10 \le p < 100, u < 10) \\ op0u & (p \ge 100, u < 10) \\ o00pu & (p < 10, u \ge 10) \\ o0pu & (10 \le p < 100, u \ge 10) \\ opu & (p \ge 100, u \ge 10) \end{cases}. \quad (3.61)$$

The 2nd-level World Grid Square code (eight numeric digits) can be described as

$$\text{2nd-level World Grid Square code} = \begin{cases} o00p0uqv & (p < 10, u < 10) \\ o0p0uqv & (10 \le p < 100, u < 10) \\ op0uqv & (p \ge 100, u < 10) \\ o00puqv & (p < 10, u \ge 10) \\ o0puqv & (10 \le p < 100, u \ge 10) \\ opuqv & (p \ge 100, u \ge 10) \end{cases}. \quad (3.62)$$

The 3rd-level World Grid Square code (10 numeric digits) can be described as

$$\text{3rd-level World Grid Square code} = \begin{cases} o00p0uqvrw & (p < 10, u < 10) \\ o0p0uqvrw & (10 \le p < 100, u < 10) \\ op0uqvrw & (p \ge 100, u < 10) \\ o00puqvrw & (p < 10, u \ge 10) \\ o0puqvrw & (10 \le p < 100, u \ge 10) \\ opuqvrw & (p \ge 100, u \ge 10) \end{cases}. \quad (3.63)$$

The 4th-level World Grid Square code (11 numeric digits) is computed as

$$\text{4th-level World Grid Square code} = \begin{cases} o00p0uqvrws_2 & (p < 10, u < 10) \\ o0p0uqvrws_2 & (10 \le p < 100, u < 10) \\ op0uqvrws_2 & (p \ge 100, u < 10) \\ o00puqvrws_2 & (p < 10, u \ge 10) \\ o0puqvrws_2 & (10 \le p < 100, u \ge 10) \\ opuqvrws_2 & (p \ge 100, u \ge 10) \end{cases}. \quad (3.64)$$

The 5th-level World Grid Square code (12 numeric digits) is computed as

$$
\text{5th-level World Grid Square code} = \begin{cases} o00p0uqvrws_2s_4 & (p < 10, u < 10) \\ o0p0uqvrws_2s_4 & (10 \leq p < 100, u < 10) \\ op0uqvrws_2s_4 & (p \geq 100, u < 10) \\ o00puqvrws_2s_4 & (p < 10, u \geq 10) \\ o0puqvrws_2s_4 & (10 \leq p < 100, u \geq 10) \\ opuqvrws_2s_4 & (p \geq 100, u \geq 10) \end{cases}, \tag{3.65}
$$

and the 6th-level World Grid Square code (13 numeric digits) is computed as

$$
\text{6th-level World Grid Square code} = \begin{cases} o00p0uqvrws_2s_4s_8 & (p < 10, u < 10) \\ o0p0uqvrws_2s_4s_8 & (10 \leq p < 100, u < 10) \\ op0uqvrws_2s_4s_8 & (p \geq 100, u < 10) \\ o00puqvrws_2s_4s_8 & (p < 10, u \geq 10) \\ o0puqvrws_2s_4s_8 & (10 \leq p < 100, u \geq 10) \\ opuqvrws_2s_4s_8 & (p \geq 100, u \geq 10) \end{cases}. \tag{3.66}
$$

As shown in Fig. 3.4, the World Grid Square coding system has a higher compatibility with the Japanese National Grid Square coding system (JIS X0410:2002). The World Grid Square coding system includes six levels of spatial resolution and enables identification of each World Grid Square by unique numeric digits (World Grid Square code) similar to JIS X0410:2002.

3.4.2 Computational Aspect of World Grid Square Statistics

Table 3.2 shows fundamental features of World Grid Squares for each level. Size of Grid Squares becomes smaller from 1st level to 6th level. There is a hierarchical structure of Grid resolutions.

The total number of 1st-level World Grid Squares over the world is $360 \times 180 \times 3/2 = 97{,}200$. The total number of 2nd-level World Grid Squares is 64 times larger than the total number of 1st level World Grid Squares. Thus, the total number of 2nd-level World Grid Squares over the world is 6,220,800. The total number of 3rd-level World Grid Squares is 100 times larger than the total number of 2nd-level World Grid Squares and it is 622,080,000. The total number of 4th-level World Grid Squares is 4 times larger than the total number of 3rd-level grid squares and it is 2,488,320,000. The total number of 5th-level Grid Squares is 4 times larger than the total number of 4th-level Grid Squares and it is 9,953,280,000. The total number of 6th-level Grid Squares is 4 times larger than 5th-level Grid Squares and it is 39,813,120,000.

Moreover, we can also approximate the total number of World Grid Squares inland since Earth's total landmass is 29.1998% of its total surface. Thus, the total number of each level World Grid Squares over the world can be estimated. Table 3.3 shows

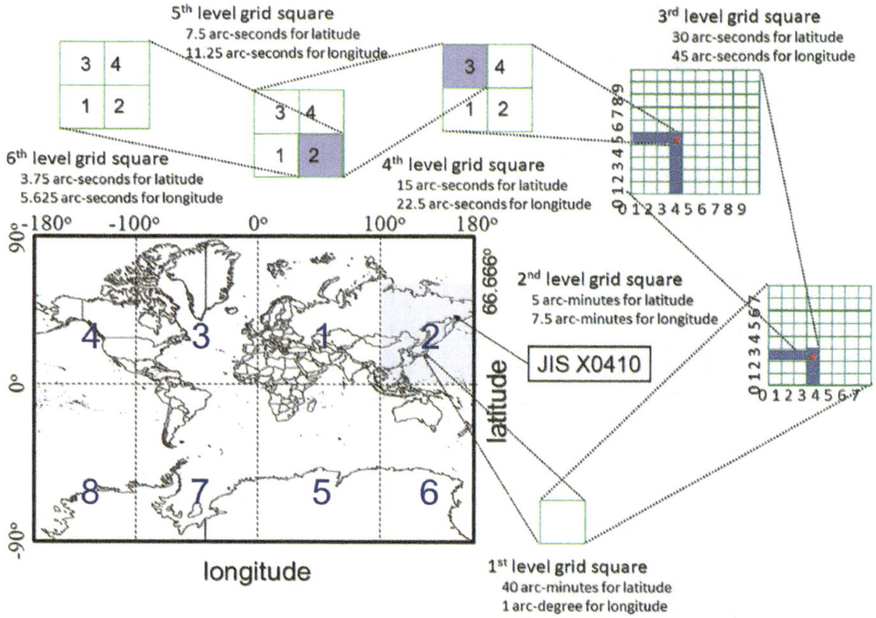

Fig. 3.4 Conceptual illustration of world grid square codes. These codes are expressed by numeric digits in which length corresponds to spatial resolution. This figure is taken from [14]

Table 3.2 Description of the World Grid Square coding system for six levels of spatial resolution.

Level	Span for latitude	Span for longitude	Length of codes
1st level	40 arc-minutes	1 arc-degree	6 numeric digits
2nd level	5 arc-minutes	7.5 arc-minutes	8 numeric digits
3rd level	30 arc-seconds	45 arc-seconds	10 numeric digits
4th level	15 arc-seconds	22.5 arc-seconds	11 numeric digits
5th level	7.5 arc-seconds	11.25 arc-seconds	12 numeric digits
6th level	3.75 arc-seconds	5.625 arc-seconds	13 numeric digits

the total number of Grid Squares and the number of Grid Squares in land for each level.

No Grid Square overlaps any other Grid Square, and each Grid Square code (Grid identification code) can be calculated from geographic information (latitude and longitude). Codes are generated by each Grid Square's geographic position on the globe. Global geodetic datum can identify a grid cell and allocate a unique Gird Square code from a geographical location. World Geodetic System 1984 (WGS84) is the most convenient geodetic datum for many users because GPS adopts WGS84 [15].

Since the World Grid Square code is highly compatible with the Japanese National Grid Square code, we can create a sequence of numeric digits to express a Grid Square

Table 3.3 Total number of Grid Squares for six levels of spatial resolution.

Level type	# of grid squares	# of grid squares in land (approximation)
1st level	97,200	28,383
2nd level	6,220,080	1,816,461
3rd level	622,080,000	181,646,116
4th level	2,488,320,000	726,584,463
5th level	9,953,280,000	2,906,337,853
6th level	39,813,120,000	11,625,351,414

Table 3.4 Description of the world grid square coding system with high spatial resolution and the total number of grid squares

Level (m)	Span for latitude	Span for longitude	Length of codes	# of grid squares	# of grid squares in land (approximation)
Extended 100	3 arc-seconds	4.5 arc-seconds	12 or 13 numeric digits	62,208,000,000	21,304,255,508
Extended 50	4.5 arc-seconds	2.25 arc-seconds	13 or 14 numeric digits	248,832,000,000	85,217,022,033
Extended 10	0.3 arc-seconds	0.45 arc-seconds	14 numeric digits	6,220,800,000,000	2,130,425,550,816
Extended 1	0.03 arc-seconds	0.045 arc-seconds	16 numeric digits	622,080,000,000,000	213,042,555,081,596

by adding two numeric digits to the upper side of the Japanese National Grid Square code.

We also mention some possible extended World Grid Square codes with high spatial resolution. Table 3.4 shows features of extended Grid Square codes with high spatial resolution and the total number of Grid Squares at each level.

3.4.3 Computer Software

Open libraries for computing World Grid Square codes with five computer languages (R, JavaScript, PHP, Java, and Python) have been released via the web page of the Research Institute for World Grid Squares [16]. There are three types of functions included in the open libraries.

Type 1: Functions that calculate the representative geographical position(s) (latitude and longitude) of a Grid Square from a given Grid Square code

Type 2: Functions that calculate a World Grid Square code from a given geographical position (latitude and longitude)

Type 3: Functions that calculate geodesic distance and size of World Grid Square (representative lengths and area) from a given Grid Square code

For example, `meshcode_to_latlong_grid()` is a function that calculates representative geographical positions (northern western and southern eastern geographic positions) from the Grid Square code, which is categorized in Type 1.

`cal_meshcode()` is a function to calculate a basic (1 km) Grid Square code (10 digits) from a geographical position, which is categorized in Type 2.

As functions categorized in Type 3, `Vincenty()` computes geodetic distance between two points placed on the WGS84 Earth ellipsoid based on Vincenty's formulae (1975) [23]. Moreover, `cal_area_from_meshcode()` calculates a size of World Grid Square code.

The following list shows functions implemented in the open libraries for translating World Grid Square code `meshcode` and geographical information (`latitude` and `longitude`) and computing spatial characteristics:

- Type 1
  - `meshcode_to_latlong(meshcode, extension = F, spec="")` calculates north-western geographic position of the grid (latitude, longitude) from `meshcode`. When you use extended Grid Square code (ex100m_12, ex100m_13, ex50m_13, ex50m_14, ex10m_14, and ex1m_16), you need to set extension = T, otherwise `extension` = F (default). Specifically, if you want to use ex100m_13 and ex50m_14, you need to set the `spec` option as "ex100m_13" and "ex50m_14", respectively.
  - `meshcode_to_latlong_NW(meshcode, extension = F, spec="")` calculates north-western geographic position of the grid (latitude, longitude) from `meshcode`. When you use extended Grid Square code (ex100m_12, ex100m_13, ex50m_13, ex50m_14, ex10m_14, and ex1m_16), you need to set extension = T, otherwise `extension` = F (default). Specifically, if you want to use ex100m_13 and ex50m_14, you need to set the `spec` option as "ex100m_13" and "ex50m_14", respectively.
  - `meshcode_to_latlong_SW(meshcode, extension = F, spec="")` calculates the geographic position (latitude, longitude) at the south-western corner of the grid with `meshcode`. When you use extended Grid Square code (ex100m_12, ex100m_13, ex50m_13, ex50m_14, ex10m_14, and ex1m_16), you need to set `extension` = T, otherwise extension = F (default). Specifically, if you want to use ex100m_13 and ex50m_14, you need to set the `spec` option as "ex100m_13" and "ex50m_14", respectively.
  - `meshcode_to_latlong_NE(meshcode, extension = F, spec="")` calculates the geographic position (latitude, longitude) at the north-eastern corner of the grid with `meshcode`. When you use extended Grid Square code (ex100m_12, ex100m_13, ex50m_13, ex50m_14, ex10m_14, and ex1m_16), you need to set `extension` = T, otherwise extension = F (default). Specifically, if you want to use ex100m_13 and ex50m_14, you need to set the `spec` option as "ex100m_13" and "ex50m_14", respectively.
  - `meshcode_to_latlong_SE(meshcode, extension = F, spec="")` calculates the geographic position (latitude, longitude) at the south-eastern corner of the Grid with `meshcode`. When you use extended Grid Square code

(ex100m_12, ex100m_13, ex50m_13, ex50m_14, ex10m_14, and ex1m_16), you need to set `extension` = T, otherwise extension = F (default). Specifically, if you want to use ex100m_13 and ex50m_14, you need to set the `spec` option as "ex100m_13" and "ex50m_14", respectively.

- `meshcode_to_latlong_grid(meshcode, extension = F, spec="")` calculates the geographic positions (latitude0, longitude0, latitude1, longitude1) at the north-western and south-eastern corner of the Grid with `meshcode`. When you use extended Grid Square code (ex100m_12, ex100m_13, ex50m_13, ex50m_14, ex10m_14, and ex1m_16), you need to set `extension` = T, otherwise extension = F (default). Specifically, if you want to use ex100m_13 and ex50m_14, you need to set the `spec` option as "ex100m_13" and "ex50m_14", respectively.

• Type 2

- `cal_meshcode(latitude,longitude)` calculates a basic (third-level) Grid Square code (10 digits) from a geographical position (`latitude, longitude`).
- `cal_meshcode1(latitude,longitude)` calculates a first-level Grid Square code (6 digits) from a geographical position (`latitude, longitude`).
- `cal_meshcode2(latitude,longitude)` calculates a second-level Grid Square code (8 digits) from a geographical position (`latitude, longitude`).
- `cal_meshcode3(latitude,longitude)` calculates a third-level Grid Square code (10 digits) from a geographical position (`latitude, longitude`).
- `cal_meshcode4(latitude,longitude)` calculates a fourth-level Grid Square code (11 digits) from a geographical position (`latitude, longitude`).
- `cal_meshcode5(latitude,longitude)` calculates a fifth-level Grid Square code (12 digits) from a geographical position (`latitude, longitude`).
- `cal_meshcode6(latitude,longitude)` calculates a sixth-level Grid Square code (13 digits) from a geographical position (`latitude, longitude`).
- `cal_meshcode_ex100m_12(latitude,longitude)` calculates an extended 100m (1 km/10) Grid Square code (12 digits) from a geographical position (`latitude, longitude`). The Grid size is 3 arc-second for latitude and 4.5 arc-second for longitude.
- `cal_meshcode_ex100m_13(latitude,longitude)` calculates an extended 100m (500 m/5) Grid Square code (13 digits) from a geographical position (`latitude, longitude`). The Grid size is 3 arc-second for latitude and 4.5 arc-second for longitude.
- `cal_meshcode_ex50m_13(latitude,longitude)` calculates an extended 50m (100 m/2) Grid Square code (13 digits) from a geographical posi-

tion (latitude, longitude). The Grid size is 1.5 arc-second for latitude and 2.25 arc-second for longitude.
- cal_meshcode_ex50m_14(latitude,longitude) calculates an extended 50m (100 m/2) Grid Square code (14 digits) from a geographical position (latitude, longitude). The Grid size is 1.5 arc-second for latitude and 2.25 arc-second for longitude.
- cal_meshcode_ex10m_14(latitude,longitude) calculates an extended 10m (100 m/10) Grid Square code (14 digits) from a geographical position (latitude, longitude). The Grid size is 0.3 arc-second for latitude and 0.45 arc-second for longitude.
- cal_meshcode_ex1m_16(latitude,longitude) calculates an extended 1m (10 m/10) Grid Square code (16 digits) from a geographical position (latitude, longitude). The Grid size is 0.03 arc-second for latitude and 0.045 arc-second for longitude.

- Type 3
  - Vincenty(latitude1, longitude1, latitude2, longitude2) calculates geodesic distance between two point indicated by (latitude1, longitude1) and (latitude2, longitude2) based on WGS84 by using the Vincenty inverse formula (1975).
  - cal_area_from_meshcode(meshcode, extension=F, spec="") calculates a size (northern west-to-east span H1, southern west-to-east span H2, north-to-south span W, and a trapezoidal approximation area A) of World Grid Square indicated by World Grid Square code meshcode. When you use extended Grid Square code (ex100m_12, ex100m_13, ex50m_13, ex50m_14, ex10m_14, and ex1m_16), you need to set extension = T, otherwise extension = F (default). Specifically, if you want to use ex100m_13 and ex50m_14, you need to set the spec option as "ex100m_13" and "ex50m_14", respectively.
  - cal_area_from_latlong(latlong) calculates a size (northern west-to-east span H1, southern west-to-east span H2, north-to-south span W, and a trapezoidal approximation area A) of a trapezoidal area indicated by latlong.

3.4.4 Examples of World Grid Square Statistics

Adopting the method provided by the Statistics Bureau of Japan, Ministry of Internal Affairs and Communications for creating Grid Square data for administrative areas in Japan [17], we produced World Grid Square data for administrative areas in more than 252 countries, using a project web page from the Research Institute for World Grid Squares [18]. Figure 3.5 shows examples of World Grid Square data for administrative

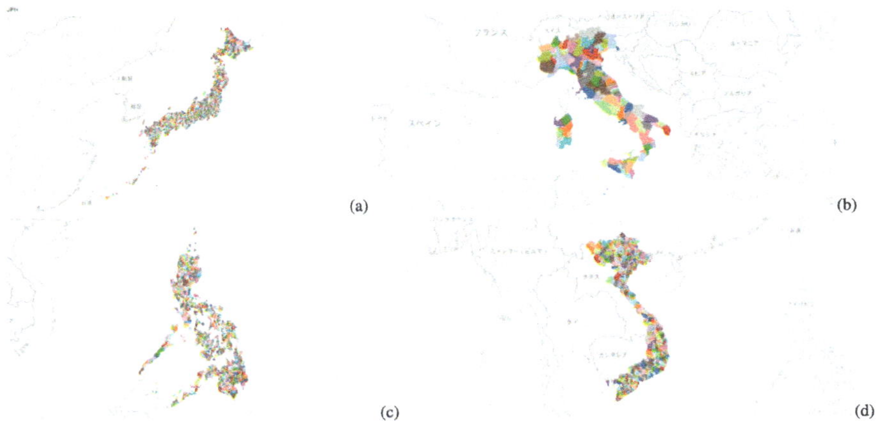

Fig. 3.5 Examples of world grid square data for administrative areas in **a** Japan, **b** Italy, **c** the Philippines, and **d** Vietnam. Grid squares included in the same administrative area are shown in the same color

areas for three countries. Grid squares included completely or partially in the same administrative area are indicated by the same color.

Additionally, World Grid Square statistics for elevations (based on satellite data) provided by the Japan Aerospace Exploration Agency (JAXA) and World Grid Square statistics for night-time light intensity provided by NASA have been released through our project page [18]. The location, extent and brightness of night-time lights can be used as primary data for measuring human artificial activities [19–21]. Figure 3.6 shows examples of World Grid Square statistics for elevation in Philippines. One record includes minimum, median, mean, and maximum elevations in a Grid Square. Furthermore, Global Soil Information can be used for creating Grid Square statistics about selected soil properties [22]. It is possible to produce various types of World Grid Square statistics from quantitative and qualitative data with geographical information.

3.5 Conclusions

This chapter explained several Grid Square coding system types: the Ordnance Survey National Grid, the European Reference Grids, and the Japanese National Grid Squares. Specifically, we introduced the Japanese National Grid Square coding system standardized in JIS X0410:2002 and the World Grid Square coding system for worldwide usage as its extension. We proposed the World Grid Square coding system and showed definitions of World Grid Square codes.

Some types of Grid Square coding systems coexist in the world. Since the Japanese National Grid Squares are based on latitude and longitude, they can extend for world-

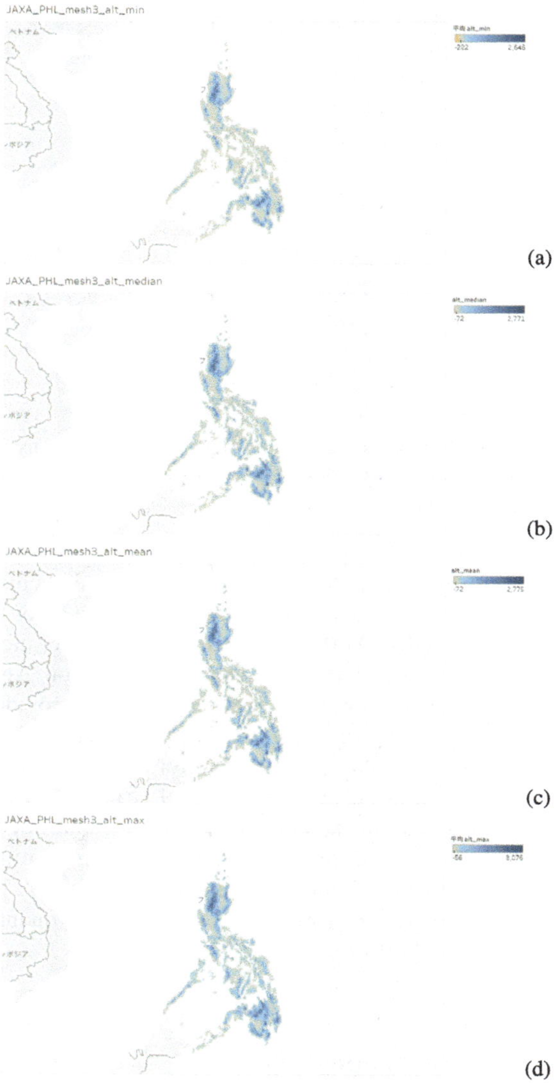

Fig. 3.6 Examples of world grid square statistics for elevations in the Philippines: **a** minimum elevations, **b** median elevations, **c** mean elevations, and **d** maximum elevations. Grid square statistics for elevations were generated from Daichi (ALOS) 30-m elevation DEM data produced by JAXA

wide usage. Other Grid Square coding systems have equi-area properties; however, a limited applicable area exists. From the extensional point of view, the Japanese National Grid Squares are practical.

We introduced a compatible extension of the Japanese National Grid Square coding system as the World Grid Square coding system and showed properties of the

World Grid Squares and computer software (R, Python, PHP, JavaScript, Java) for World Grid Squares. The Research Institute provides three types of functions for World Grid Squares as open libraries. A World Grid Square code is computed from given geographical information (latitude and longitude), and representative locations of World Grid Squares are calculated from given World Grid Square codes. The area and size of World Grid Squares are also computable in open libraries.

Moreover, we mentioned some examples of World Grid Square data and statistics using polygon data about administrative areas and gridded satellite data.

References

1. J.G. Granö, *Pure Geography*, ed. by O. Grano, A. Paasi, M. Hicks (Johns Hopkins University Press, Baltimore, MD, 1997)
2. Ordnance Survey. https://www.ordnancesurvey.co.uk/support/the-national-grid.html. Accessed 21 June 2014
3. JISC, The Japanese National Grid Square Codes, JIS X0410:2002
4. Eurostat, Grid Maker. https://github.com/eurostat/GridMaker. Accessed 13 Jan 2024
5. https://sgis.kostat.go.kr/view/map/interactiveMapMain . Accessed 10 Dec 2023
6. National Nested Grid. https://www.anzlic.gov.au/resources/national-nested-grid. Accessed 13 Jan 2024
7. GEOSTAT 1A. http://www.efgs.info/geostat/1a. Accessed 14 Sept 2015
8. GEOSTAT 1B. https://www.efgs.info/geostat/1b. Accessed 24 Mar 2022
9. Australian Bureau of Statistics, Australian Bureau of Statistics, Regional population, Reference period 2021–22 financial year. https://www.abs.gov.au/statistics/people/population/regional-population/2021-22. Accessed 7 Jan 2024
10. UNECE, Generic Statistical Business Process Model. https://statswiki.unece.org/display/GSBPM/GSBPM+v5.0. Accessed 12 Dec 2023
11. OGC Discrete Global Grid System (DGGS) Core Standard (15-104r3). http://portal.opengeospatial.org/files/66643. Accessed 10 Nov 2016
12. Uber Technologies, Inc., Hexagonal hierarchical geospatial indexing system. https://h3geo.org. Accessed 10 Jan 2024
13. Research Institute for World Grid Squares. https://www.fttsus.jp/worldgrids/en/library/. Accessed 24 Dec 2023
14. A.-H. Sato, S. Nishimura, H. Tsubaki, World Grid Square codes: definition and an example of World Grid Square data, in *2017 IEEE International Conference on Big Data (BIGDATA)*, Dec 11–14 (2017), pp. 4156–4165
15. Department of Defense World Geodetic System 1984: Its definition and relationships with local geodetic systems, National Imagery and Mapping Agency. TR8350.2. 3 Jan 2000
16. Research Institute for World Grid Squares. http://www.fttsus.jp/worldgrids/en/library. Accessed 28 Apr 2018
17. A list of grid square codes for 2nd level administrative areas in Japan, Statistics Bureau, Ministry of Internal Affairs and Communications. http://www.stat.go.jp/data/mesh/m_itiran.htm
18. Research Institute for World Grid Squares. http://www.fttsus.jp/worldgrids/
19. C.D. Elvidge, P. Cinzano, D.R. Pettit, J. Arvesen, P. Sutton, C. Small, R. Nemani, T. Longcore, C. Rich, J. Safran, J. Weeks, S. Ebener, The Nightsat mission concept. Int. J. Remote Sens. **28**(12), 2645–2670 (2007)
20. B. Klemens, A. Coppola, M. Sbron, Estimating Local Poverty Measures Using Satellite Images, World Bank Group, Policy Research Working Paper, 7329 (2015)
21. C. Mellander, J. Lobo, K. Stolarick, Z. Matheson, Night-time light data: a good proxy measure for economic activity? PLoS ONE **10**(10), e0139779 (2015)

22. T. Hengl, J.M. de Jesus, R.A. MacMillan, N.H. Batjes, G.B.M. Heuvelink et al., SoilGrids1km—global soil information based on automated mapping. PLoS ONE **9**(8), e105992 (2014)
23. T. Vincenty, Direct and inverse solutions of geodesics on the ellipsoid with application of nested equations. Surv. Rev. XXIII **176**, 88–93 (1975)

Chapter 4
Statistical Foundations About Grid Square Statistics

Abstract This chapter introduces the statistical properties of Grid Square statistics. Grid square statistics can be described as tabular expressions, especially as matrix data or a contingency table in case of summarization for frequency or categorical data. We explain how to produce Grid Square statistics from qualitative and quantitative data. We further discuss errors in Grid Square statistics created as a statistical inference based on random sampling. For example, there are several Grid Square statistics about the number of persons within a Grid Square based on logs of mobile communication such as mobile phones and Wi-Fi base stations and land cover categorization produced by satellite images. Moreover, we confirm basic arithmetic operations between Grid Square statistics and explain the most fundamental procedure to convert Grid Square statistics on one Grid coordinate system into Grid Square statistics on another Grid coordinate system with concrete examples.

4.1 Applications of Grid Square Statistics

In 1960, Statistics Bureau of Japan began to study the definition of Grid Squares for computing spatial statistics from a horizontal perspective across other Japanese government ministries and departments. In the 1970s, they proposed Grid Square codes computed on the basis of geographical position (latitude and longitude), and this definition was adopted as Japanese Industrial Standards in 1976 (JIS C 6304:1976). Adopting this definition, Japanese government ministries and departments and Japanese firms in industrial sectors have created Grid Square statistics for the purposes of planning, assessment, and evaluation in policy-making.

During more than 40 years of use in Japan, Japanese National Grid Square codes and various types of Grid Square statistics have been provided by the government and by industrial organizations [1]. As one example, the Statistics Bureau of Japan, Ministry of Internal Affairs and Communications, provides a statistical GIS service supplying Grid Square statistics for population and economic census data [2]. Similarly, the National Land Information Division of the National Spatial Planning and Regional Policy Bureau (Ministry of Land, Infrastructure, Transport, and Tourism)

A.-H. Sato and H. Tsubaki, *Evaluation Platform of Sustainability for Global Systems*,
https://doi.org/10.1007/978-981-97-2296-9_4

provides public access to Grid Square statistics for numerical land information, future population projections, tourism [3].

Grid Square statistics enable two different Grid Square statistics to be merged and integrated, and they can be retotaled following a given area. Because Grid Square statistics are produced by qualitative or quantitative data with geographical information (latitude and longitude), they have the advantage of not being influenced by temporal changes of administrative area.

Since Japan adopted the World Geodetic System (ISO 6709) in 2002, all official geodetic measurements must be scaled in this way, with a 10-year moratorium. As a result, Japan's National Grid Square coding system was also revised to conform with the World Geodetic System (JIS X0410:2002), and since 2012, all Grid Square statistics and data have been produced on that basis. In this sense, the Japanese National Grid Square coding system is theoretically applicable to countries other than Japan.

Six types of applications to Grid Square statistics are well-known:

- Data linkage and data processing
- Mapping
- Data curation on given areas
- Identifying effective areas
- Defining observation areas
- Units for numerical simulation.

In the data linkage and data processing application, we can link different Grid Square statistics and use them in operations among other Grid Square statistics to synthesize new Grid Square statistics from several original data sources. In the mapping application, we can visualize Grid Square statistics on a map to analyze the focal area.

In the data curation on given area applications, we can produce statistics on a given area by recalculating Grid Square statistics for that area. In identifying effective area applications, Grid Square statistics are used regarding distances and areas. In applications to define observation areas, Grid Square statistics can be used to determine an area for the collection of data or samples. Finally, in the numerical simulation application, Grid Square statistics can conduct numerical simulations for a unit, such as diffusion processes, percolation models, and migration processes.

These are examples of applications for recalculating Grid Square statistics and mapping Grid Square statistics about population, land consumption, green cover rates, and so on.

Moreover, if we use the same standard Grid Square coordinate system, Grid Square statistics produced by one person or organization can connect with Grid Square Statistics produced by another person or organization. Such a principle enables us to conduct aggregation and correlation analysis among Grid Square Statistics. Some examples are shown in Chap. 8.

4.2 Tabular Representation of Grid Square Statistics

In case of summarization for frequency or categorical data, Grid Square statistics are expressed as a contingency table consisting of a unique key (Grid Square code) representing Grid Square and its statistics about the number of categorical data.

Figure 4.1 shows an example to produce Grid Square statistics from point data of persons and objects. Typically, the production of Grid Square statistics requires the existence of geographically referred point data with high spatial accuracy. For example, in most cases, geocoding can transform addresses registered for business purposes into geographic coordinates. Then, in a second step, these points of interest (POI) can be aggregated by Grid Squares (See Fig. 4.1).

For this purpose, the Japanese National Grid Square coding system standardized as JIS X0410:2002 has a long history of being used for high-resolution government statistics developed in the 1970s.

Other Grid Square statistics can be produced from gridded data with a spatial resolution higher than Grid Squares used as aggregation units. An example of this type is Grid Square statistics produced from satellite images. One example of such Grid Square statistics is about nighttime light intensity produced from satellite images recorded as earth observation data [4].

It is essential to understand data types to produce and analyze statistics. According to the general definition of data, data is classified into two types: qualitative and quantitative. The qualitative data is expressed using letters and characters, such as words, unique numeric sequences, and labels. Qualitative data is also classified into nominal scale and ordinal scale. Quantitative data is expressed using numbers. Quantitative data is also classified into interval scale and ratio scale. This classification is shown as follows:

- Qualitative data
  - Nominal scale
  - Ordinal scale.

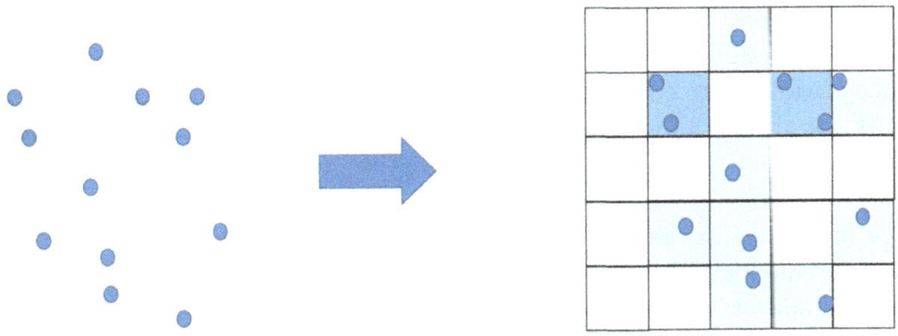

Fig. 4.1 An example of grid square statistics produced from point data

- Quantitative data
  - Interval scale
  - Ratio scale.

Statistics is computed from qualitative data as frequency by counting the number of nominal scale data which satisfy some conditions. Statistics of quantitative data are some representative values such as mean, median, and order statistics.

This fundamental property of statistics also applies to Grid Square Statistics, and the general principle of statistics for qualitative and quantitative data is fulfilled. In qualitative data, Grid Square Statistics can be expressed as a matrix data between qualitative data and Grid Square code. In the case of quantitative data, they can be described as a table between representative values and Grid Square code.

Table 4.1 shows examples of Grid Square Statistics for qualitative and quantitative data. Table 4.1a shows Grid square statistics computed from two types of qualitative data. Grid Square statistics are categorized into two types depending on data types. This table shows Grid Square statistics by expressing three times r matrix tabular data for two categories of data, including Grid Square code. In this sense, Grid Square statistics are not geographical data but statistics, including Grid Square codes expressed by tables (matrix).

Table 4.1b shows Gird Square statistics from quantitative data. This table includes the mean temperature and elevation of Grid Square codes. Since temperature and elevation are quantitative data, the mean is used as a representative value. Representative values such as median, variance, and standard deviations are acceptable.

4.2.1 *How to Produce Grid Square Statistics*

A method to produce Grid Square statistics depends on the data type (qualitative and quantitative). Grid Square statistics for qualitative data is used as frequency by counting the number of data of which some conditions are satisfied. Grid Square statistics for quantitative data are some representative values such as mean, median, and order statistics.

Grid Square statistics are produced based on a Grid with a unique numeric sequence (Grid Square code) and a tiny rectangular region in a given domain D.

Let W_i be the ith Grid Square expressed by Grid Square code w_i. Since Grid Squares are mutually exclusive, we have $W_i \cap W_j = \phi$ for any i and j $(i \neq j)$ and $\cup_{i=1}^{M} W_i = D$.

If we have physical address data for customers with sample size n, we produce Grid Square statistics for the number of customers in terms of Grid Square code since the physical address is converted to Grid Square code. Table 4.2 shows Grid Square statistics for the number of customers and relative frequency. Of course, we have

Table 4.1 **a** The tabular expression of Grid Square statistics of two types of categorical data (the number of person and the number of object A) by expressing a $3 \times r$ matrix. **b** Tabular expression for Grid Square statistic of mean temperature and mean elevation (interval scale data) by expressing a $3 \times r$ matrix

(a)

Grid square code	Population	The number of object A
2054111214	45	4
2054111215	341	5
2054111216	3413	76
2054111217	415	1
2054111218	315	34
2054111219	641	14
2054111220	24	1

(b)

Grid square code	Mean temperature (°C)	Mean elevation (m)
2054111214	25.2	424.5
2054111215	34.1	25.5
2054111216	34.3	76.4
2054111217	21.5	154.2
2054111218	31.5	234.4
2054111219	25.1	145.5
2054111220	24.5	132.5

Table 4.2 Grid Square statistics for the number of customers

Grid square code	The number of customers	Relative frequency
w_1	d_1	d_1/n
w_2	d_2	d_2/n
w_3	d_3	d_3/n
w_4	d_4	d_3/n
...	...	...
w_m	d_m	d_m/n
Total	n	1

$$\sum_{k=1}^{m} d_k = n, \tag{4.1}$$

$$\sum_{k=1}^{m} d_k/n = 1. \tag{4.2}$$

Table 4.3 Record data consisting of postal code, gender (M or F), height, and weight

Postal code	Gender (M or F)	Height (cm)
21112	M	175
21112	F	170
21112	M	171
21112	M	172
21113	F	171
21113	F	170
21113	M	165

Table 4.4 Record data consisting of postal code, gender (M or F), height, and weight

Grid square code	Population	Male population	Female population	Mean height (cm)	Minimum height (cm)	Maximum height (cm)
2052121111	4	3	1	172	170	175
2052121112	3	2	1	168.6667	165	171

Therefore, we can determine the Grid Square code for location information included in the data. Various data collection types in business activities contain location information such as address, postal code, and latitude and longitude. However, location information is easily connected to personal information. It is not easy to analyze location information directly under privacy protection law.

For example, as shown in Table 4.3, we have record data consisting of postal code (location information), gender (male or female), height, and weight.

Since the volume of postal code is geographical information we can convert it into Grid Square code approximately. Since the gender column is of a nominal scale in qualitative data, the frequency of two categories and their total is composed as Grid Square statistics. Moreover, the height and weight columns are of ratio scale in quantitative data. Thus, representative values are computed as Grid Square statistics.

Therefore, we can compute Grid Square statistics such as population, male population, female population, height (sample mean, minimum, maximum), and representative values (sample mean, minimum, maximum) for weight. If we identify the Grid Square code of postal codes 21112 and 21113 as 2052121111 and 2052121112 (approximately), then we compute Grid Square statistics as shown in Table 4.4.

4.3 Statistical Inference

4.3.1 Errors of Grid Square Statistics Produced from the Random Samples of Categorical Data

Errors of Grid Square statistics produced from random samples can be measured by different framework in terms of categorical data such as ordinal or nominal scale. Error estimation of Grid Square statistics produced from samples for categorical data is based on binomial distribution or normal distribution. Error estimation of Grid Square statistics represented by mean produced from samples for categorical data is based on approximation of population mean by sample mean.

For a sake of simplicity, we suppose a binary classification problem. Suppose that two classes (A ($X = 1$) and B($X = 0$)) expressed by a Bernoulli random variable $X = 1 \quad (w.p.\ p)$ and $X = 0 \quad (w.p.\ 1 - p)$.

Assume that m samples are observed as class A and $n - m$ samples are as class B from n samples in a Grid Square. The sampling ratio is calculated as $\tilde{p} = m/n$ taking $X = 1$. Since m follows binomial distribution $B(n, p)$, m is approximately sampled from normal distribution $N(np, np(1 - p))$:

$$\tilde{p} - 1.96\sqrt{\frac{\tilde{p}(1-\tilde{p})}{n}} \leq p \leq \tilde{p} + 1.96\sqrt{\frac{\tilde{p}(1-\tilde{p})}{n}}. \tag{4.3}$$

When the population size Pop_0, sample size n by random sampling population, and $m(w)$ samples are observed in a grid square w, the estimates of the number of objects found in a Grid Square w and their 95% confidence interval are approximated as

$$Pop_0\left(\tilde{p}(w) - 1.96\sqrt{\frac{\tilde{p}(w)(1-\tilde{p}(w))}{n}}\right) \leq Pop(w) \leq Pop_0\left(\tilde{p}(w) + 1.96\sqrt{\frac{\tilde{p}(w)(1-\tilde{p}(w))}{n}}\right). \tag{4.4}$$

where we set $\tilde{p}(w) = m(w)/n$.

Errors of Grid Square statistics produced from the random samples of categorical data such as nominal or ordinal scale is calculated as an approximation of population mean by sample mean. Suppose that n ordinal or nominal variables $x_i \quad (i = 1, 2, \ldots, n)$ and estimate population mean μ from sample mean $\tilde{\mu}$. The sample mean is calculated as

$$\tilde{\mu} = \frac{1}{n}\sum_{i=1}^{n} x_i. \tag{4.5}$$

Since $\tilde{\mu}$ are approximately sampled from normal distribution $N(\mu, \sigma^2/n)$ for sufficiently large n, where $\sigma^2 = \frac{1}{n}\sum_{i=1}^{n}(x_i - \tilde{\mu})^2$. Thus, 95% confidence interval of population mean μ follows:

$$\tilde{\mu} - 1.96\sigma \leq \mu \leq \tilde{\mu} + 1.96\sigma. \tag{4.6}$$

4.4 A Problem to Adjust an Estimation Ratio with Some Classification Errors

Some types of Grid Square statistics inferred from categorical data are influenced by both sample and classification errors. Grid Square statistics about land cover ratios are inferred from gridded classification categorical data. They can be influenced by classification errors as well as sample errors. For example, Grid Square statistics about green-cover index requires a fractional cover algorithm to calculate the percentage of green cover, non-green cover and bare ground for each pixel. Then, the percentage of green cover and non-green cover are strictly affected by classification errors.

Suppose a binary classification problem with some classification errors. Table 4.5 shows a 2×2 confusion matrix for binary classification with a finite number of validations.

Here TP, FN, FP, and TN are the number of observations classified as follows:

- TP; a classification (c) is taken as $X = 1$ and a validation (v) is taken as $X = 1$
- FN; a classification (c) is taken as $X = 1$ and a validation (v) is taken as $X = 0$
- FP; a classification (c) is taken as $X = 0$ and a validation (v) is taken as $X = 1$
- TN; a classification (c) is taken as $X = 0$ and a validation (v) is taken as $X = 0$.

Thus, under the finite number of verification actions, the conditional probability estimates can be computed as follows:

$$\begin{aligned} R(0|0) &= \frac{TN}{FP+TN}, \ R(1|0) = \frac{FP}{FP+TN}, \\ R(0|1) &= \frac{FN}{TP+FN}, \ R(1|1) = \frac{TP}{TP+FN}. \end{aligned} \tag{4.7}$$

Then, when the population ratio taking $X = 1$ is denoted as p, the estimation ratio $\hat{p} = m/n$ with some classification errors can be expressed as

$$\hat{p} = R(1|1)p + R(1|0)(1-p). \tag{4.8}$$

Thus, the population ratio p is

Table 4.5 An example of 2×2 confusion matrix

	Validated v		
	c/v	$X = 1$	$X = 0$
Classified c	$X = 1$	TP	FP
	$X = 0$	FN	TN

$$p = \frac{\hat{p} - R(1|0)}{R(1|1) - R(1|0)}. \tag{4.9}$$

Moreover, under the finite number of verification actions, 95% confidence interval of conditional probabilities may be expressed as

$$\begin{aligned} &\frac{TN}{FP+TN} - 1.96\sqrt{\frac{TN}{FP+TN}\left(1-\frac{TN}{FP+TN}\right)/(FP+TN)} \le R(0|0) \\ &\le \frac{TN}{FP+TN} + 1.96\sqrt{\frac{TN}{FP+TN}\left(1-\frac{TN}{FP+TN}\right)/(FP+TN)}, \end{aligned} \tag{4.10}$$

$$\begin{aligned} &\frac{FP}{FP+TN} - 1.96\sqrt{\frac{FP}{FP+TN}\left(1-\frac{FP}{FP+TN}\right)/(FP+TN)} \le R(1|0) \\ &\le \frac{FP}{FP+TN} + 1.96\sqrt{\frac{FP}{FP+TN}\left(1-\frac{FP}{FP+TN}\right)/(FP+TN)}, \end{aligned} \tag{4.11}$$

$$\begin{aligned} &\frac{FN}{TP+FN} - 1.96\sqrt{\frac{FN}{TP+FN}\left(1-\frac{FN}{TP+FN}\right)/(TP+FN)} \le R(0|1) \\ &\le \frac{FN}{TP+FN} + 1.96\sqrt{\frac{FN}{TP+FN}\left(1-\frac{FN}{TP+FN}\right)/(TP+FN)}, \end{aligned} \tag{4.12}$$

$$\begin{aligned} &\frac{TP}{TP+FN} - 1.96\sqrt{\frac{TP}{TP+FN}\left(1-\frac{TP}{TP+FN}\right)/(TP+FN)} \le R(1|1) \\ &\le \frac{TP}{TP+FN} + 1.96\sqrt{\frac{TP}{TP+FN}\left(1-\frac{TP}{TP+FN}\right)/(TP+FN)}. \end{aligned} \tag{4.13}$$

Specifically, in the case that $\frac{\delta a}{a}$ and $\frac{\delta b}{b}$ are sufficiently small, error transfer formulas for four arithmetic operations are given as

$$\begin{aligned} (a \pm \delta a) + (b \pm \delta b) &= (a+b) \pm (\delta a + \delta b), \\ (a \pm \delta a) - (b \pm \delta b) &= (a-b) \pm (\delta a + \delta b), \\ (a \pm \delta a)(b \pm \delta b) &= ab\left(1 \pm \left(\frac{\delta a}{a} + \frac{\delta b}{b}\right)\right), \\ \frac{a \pm \delta a}{b \pm \delta b} &= \frac{a}{b}\left(1 \pm \left(\frac{\delta a}{a} + \frac{\delta b}{b}\right)\right). \end{aligned}$$

Therefore, we may calculate 95% confidence interval for population ratio p with classification errors as follows:

$$\frac{a-b}{c-b}\left(1-\left(\frac{\delta a+\delta b}{a-b}+\frac{\delta c+\delta b}{c-b}\right)\right) \le p \le \frac{a-b}{c-b}\left(1+\left(\frac{\delta a+\delta b}{a-b}+\frac{\delta c+\delta b}{c-b}\right)\right), \tag{4.14}$$

where we set

$$a = \hat{p}, b = R(1|0), c = R(1|1),$$

and δa, δb, and δc are a half of the 95% confidence intervals of $\hat{p}$, $R(1|0)$, and $R(1|1)$, respectively. Of course, the inequality expressed as Eq. (4.14) is valid for $\frac{\delta a+\delta b}{a-b} \ll 1$ and $\frac{\delta c+\delta b}{c-b} \ll 1$. Classification errors should be sufficiently small, namely, $R(1|0) \ll 1$. Moreover, an estimation ratio $\hat{p}$ with classification errors are sampled from a binomial distribution, which is approximated by a normal distribution.

4.5 Calculation

As a processing method that takes advantage of the characteristics of Grid Square statistics and data, we consider combining the four basic arithmetic operations in the same Grid Square. In these operations, the sum and difference must be in the same system of units. Products and ratios are calculated between different systems of units, and the resulting system of units is the product or ratio of the original two values's unit. The product operation can be used to calculate different quantities from ratios and quantities. Also, ratio arithmetic can be used to calculate proportions from different quantities. We also mention conversion procedure from one Grid Square coordinate system to another Grid Square coordinate system.

4.5.1 *Basic Arithmetic Operations*

Suppose that there are n types of Grid Square statistics data in Grid Square code i, $X_i = \{x_{i,1}, x_{i,2}, \ldots, x_{i,n}\}$ and n types of Grid Square statistics data in Grid Square code j, $X_j = \{x_{j,1}, x_{j,2}, \ldots, x_{j,n}\}$. At this time, we take two Grid Square statistics data from each of them, and consider their four arithmetic operations. These four arithmetic operations are (1) sum $x_{i,k} + x_{j,k}$, (2) difference $x_{i,k} - x_{j,k}$, (3) product $x_{i,k} \cdot x_{j,k}$, and (4) ratio $x_{i,k}/x_{j,k}$.

We show the following examples for each arithmetic operation:

(1) sum: If $x_{i,k}$ is the urban area ratio of Grid Square i and $x_{i,l}$ is the crop field ratio of Grid Square i, then $x_{i,k} + x_{i,l}$ is the area ratio used as the urban area or crop field of Grid Square i.
(2) difference: If $x_{i,k}$ is the forest area ratio of Grid Square i and $x_{i,l}$ is the occupancy rate of the coniferous forest ratio of Grid Square i, then $x_{i,k} - x_{i,l}$ is the area ratio of non-coniferous forests in Grid Square k.
(3) product: If $x_{i,k}$ is the occupancy ratio of the accommodation facilities in Grid Square i, and $x_{j,k}$ is the accommodation capacity of the accommodation facilities in Grid Square j, then $x_{i,k}x_{j,k}$ is the number of guests in Grid Square i.
(4) ratio: If $x_{i,k}$ is the amount of sales of the secondary industry in Grid Square i and $x_{i,l}$ is the number of workers in the secondary industry in Grid Square i, then $x_{i,k}/x_{i,l}$ is the amount of sales of the secondary industry per person employed in Grid Square i.

Assuming that the value of the Grid Square statistics is additive quantitative data, let $W = \{w_1, w_2, \ldots, w_N\}$ be the set of Grid Square codes having an intersection with a closed region Z. At this time, the additive Grid Square statistics x_i corresponding to the Grid Square code included in W can be added together to calculate an approximate value of the aggregate value for the closed region Z. For example, if the area within a radius of 5 km around the Grid Square of interest is Z, and x_i is the Grid Square statistic of the total population,

$$X = \sum_{i=\arg_{w_k \in W} w_k} x_i \tag{4.15}$$

is the population living within 5 km of the place of interest. Such population X is used as an estimate of the potential population that is likely to visit the store in the trade area analysis. Furthermore, by calculating the surrounding population while changing the Grid Square of interest in the center, we can search for promising locations for opening stores (store opening plan). Furthermore, by taking the area included in the administrative division (municipality) as Z, it is possible to approximate statistics for administrative area using the Grid Square statistics as the basic unit and the administrative division as the aggregation unit.

There are several methods for obtaining a set of Grid Square codes W having an intersection with a closed region Z. In preparation, we define area ratio (distribution ratio) of the area $S(Z \cap w_i)$ for the intersection $Z \cap w_i$ between the closed region Z and the Grid Square w_i to the area $S(w_i)$ for Grid Square w_i,

$$\rho(Z, w_i) = \frac{S(Z \cap w_i)}{S(w_i)}. \tag{4.16}$$

Obviously, we have

$$0 \leq \rho(Z, w_i) \leq 1. \tag{4.17}$$

For a sake of simplicity, there are three ways to calculate statistic within the closed region Z. As shown in Eq. (4.15), one is to take a set of Grid Square W_u as a Grid Square with a non-zero area ratio $\rho(Z, w_i)$. The other is to take a set of Grid Square W_l where an area ratio $\rho(Z, w_i)$ equals to 1. If the true statistic aggregated from the original data is X^* directly for region Z without using Grid Square statistics, then it is obvious that

$$\sum_{i \in \arg_{w_k \in W_l} w_k} x_i < X^* < \sum_{i \in \arg_{w_k \in W_u} w_k} x_i \tag{4.18}$$

is satisfied. Equation (4.18) gives the upper and lower limits of the statistical value. Furthermore, by calculating

$$X^{\alpha} = \sum_{i \in \arg_{w_k \in W} w_k} \rho(Z, w_i) x_i \tag{4.19}$$

using the area ratio (distribution ratio) $\rho(Z, w_i)$, X^α is a better approximation of X^* by area apportionment identification. Furthermore, by modifying the method for calculating $\rho(Z, w_i)$, we can obtain an approximate value by other identification methods.

4.5.2 Conversion Method Between Different Grid Coordinate Systems

This section explains conversion procedure from one Grid Square coordinate system to another Grid Square coordinate system. We mentioned existing Grid coordinate systems in Sect. 3.2. The following list shows several existing Grid coordinate systems that define Grid Square (grid, cell or polygons) and unique Grid Square codes to identify them.

- Japanese National Grid Square coding system [5]: JIS X0410:2002 (JISC)
- World Grid Square coding system [6]: compatible extension to JIS X0410 (RIWGS)
- European Reference Grids [7]: ETRS89-LAEA (INSPIRE, EU Commission)
- National Nested Grid (NNG) [8]: the Australia and New Zealand Land Information Council (ANZLIC)
- British National Grid (BNG) or Ordnance Survey National Grid [9]: Ordnance Survey
- Global Discrete Grid System (GDGS) [10]: ISO 19170-1:2021

One of the essential properties of Grid Square statistics is the ability to share Grid Square statistics on a standard Grid coordinate system. Data connectability based on common Grid Square codes enables us to conduct correlation and regression analyses on Grid Square statistics. However, we may lose data connectability between Grid Square statistics produced on different Grid coordinate systems. We cannot directly compare Grid Square statistics produced on different Grid coordinate systems to each other.

In this section, we will show how to approximately convert Grid Square statistics of one Grid coordinate system into those of other Grid coordinate system by using proportionally dividing values based on area ratio of one type of Grid Square to other type of Grid Square.

We consider two types of methods to choose the area ratio. Suppose that Grid Square code e and Grid Square statistics x_e on one Grid coordinate system. Let us consider converting statistics x_e into statistics y_w in Grid Square coded as w on other Grid coordination system. We define a set $Cover(e)$ of Grid Square codes on other Grid coordinate system whose Grid Squares cover Grid Square coded as e on one Grid coordination system as shown in Fig. 4.2.

Assume that a Grid Square V_e is codded as e on one Grid Square coordinate system and that a Grid Square W_w is coded as w on other Grid Square coordinate system. From statistics x_e on one Grid Square coordinate system, statistics y_w on

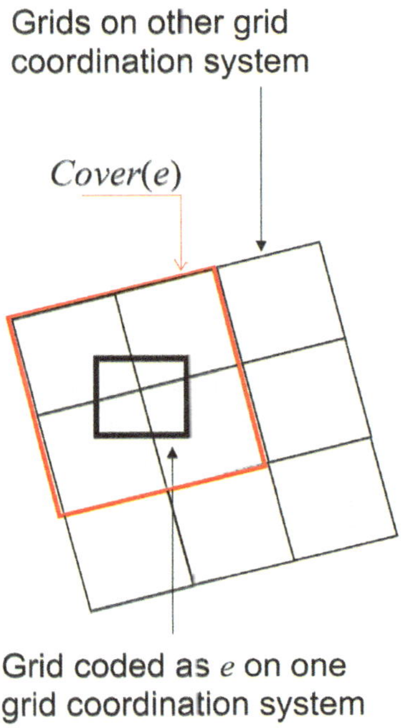

Fig. 4.2 Relationship between one grid coordinate system and other grid coordinate system

other Grid Square coordinate system is computed as

$$y_w = \sum_{e' \in Cover(w)} \rho(W_w, V'_{e'}) x_{e'}, \tag{4.20}$$

where

$$\rho(W_{w'}, V'_e) = \begin{cases} \frac{S(W_{w'} \cap V'_e)}{S(V'_e)} & w' \in Cover(e) \\ 0 & w' \notin Cover(e) \end{cases} \tag{4.21}$$

Here, $S(g)$ represents an area of Grid Square g and V'_e is a Grid Square converted into other Grid coordinate system from a Grid Square V_e on one Grid coordinate system.

We address the procedure to convert Grid Square statistics computed based on one Grid coordinate system into those on other Grid coordinate system by using proportionally dividing values based on area ratio of one type of Grid Square to other type of Grid Square.

1. Identify a Grid Square V_e on a Grid coordinate system from a Grid Square code e.
2. Convert a Grid Square V_e on one Grid coordinate system into a Grid Square V'_e on other Grid coordinate system.

3. Generate a set of Grid Square codes $Cover(e)$ for other Grid coordinate system covering a Grid Square V_e'.
4. Generate a set of Grid Squares $W_{w'}$ included in a set of Grid Square codes $Cover(e)$.
5. Compute area ratio $\rho(W_{w'}, V_e')$ from Grid Squares $W_{w'}$ and V_e'.
6. By using Eq. (4.20), compute statistical value y_w for a Grid Square W_w of other Grid coordinate system from a set of area ratios $\rho(W_{w'}, V_e')$ and a set of statistical values x_e for Grid Squares V_e included in a set of Grid Square codes $Cover(w)$.

4.5.3 Examples

In this section, we show some examples to convert Grid Square statistics into Grid Square statistics on the World Grid Square coordinate system [6] introduced in Sect. 3.4. The World Grid Square coordinate system is an extension of the Japanese National Grid Square coding system standardized as JIS X0410:2002 [5]. We explain how to convert Grid Square statistics produced on the Australian National Grid coordinate system and European Reference Grid coordinate system into Grid Square statistics on the World Grid Square coordinate system. We apply the procedure mentioned in Sect. 4.5.2 to converting Grid Square statistics.

4.5.3.1 Australian Population Grids

Let consider converting Grid Square statistics on the Australian National Grid coordinate system (National Nested Grid) [8] into Grid Square statistics on the World Grid Square coordinate system [6].

The Australian National Grid coordinate system is defined on geodesic epilsoid GRS80 and Albers equarea conic projection with center position latitude 0° and longitude 132° and standard latitude liens are $-18°$ and $-36°$. The World Grid Square coordinate system is defined on geodesic epilsoid WGS84 and latitude and longitude projection.

The conversion procedure is as follows:

1. Identify a Grid Square V_e on the Australian Reference Grid coordinate system from a Grid Square code e for Grid Square statistics x_e.
2. Convert a Grid Square V_e on the Australian Reference Grid coordinate system into a Grid Square V_e' on the World Grid Square coordinate system.
3. Generate a set of Grid Square codes $Cover(e)$ for the World Grid Square coordinate system covering a Grid Square V_e'.
4. Generate a set of Grid Squares $W_{w'}$ included in a set of Grid Square codes $Cover(e)$.
5. Compute area ratio $\rho(W_{w'}, V_e')$ from Grid Squares $W_{w'}$ and V_e'.

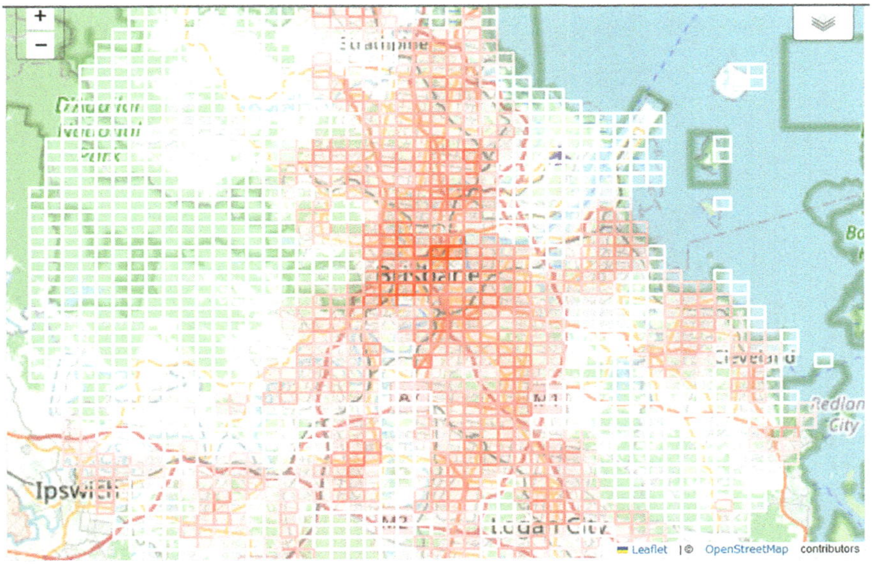

Fig. 4.3 Grid square map of Australian population grids 2016 converted into world grid square statistics around Brisbane

6. By using Eq. (4.20), compute statistical value y_w for a Grid Square W_w of the World Grid Square coordinate system from a set of area ratios $\rho(W_{w'}, V'_e)$ and a set of statistical values x_e for Grid Squares V_e included in a set of Grid Square codes $Cover(w)$.

Figure 4.3 shows Grid Square map representation of Australian population grids 2016 provided from Australian Bureau of Statistics [11] converted into World Grid Square statistics. The depth of color represents the degree of population on each World Grid Square.

4.5.3.2 European Reference Grids

We consider converting Grid Square statistics on the European Reference Grid coordinate system [7] into Grid Square statistics on the World Grid Square coordinate system [6] by applying the procedure mentioned in Sect. 4.5.2 to converting Grid Square statistics.

The European Reference Grid coordinate system is defined on geodesic epilsoid GRS80 and based on the ETRS89 Lambert Azimuthal Equal-Area Projection Coordinate Reference System (ETRS89/ETRS-LAEA), which uses a projection point set at 52°N and 10°E.

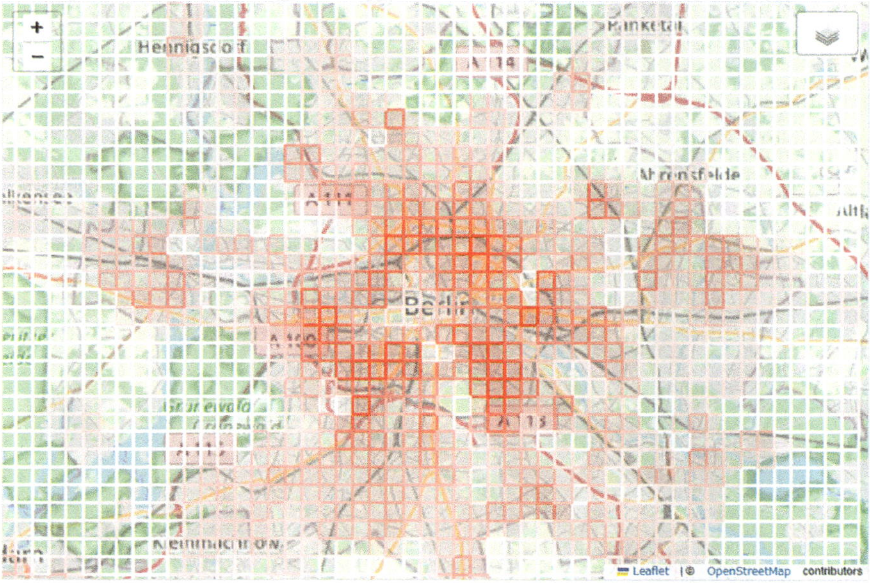

Fig. 4.4 Grid square map of GEOSTAT 1 km^2 population grid 2011 converted into world grid square statistics around Berlin, Germany

The conversion procedure is as follows:

1. Identify a Grid Square V_e on the European Reference Grid coordinate system from a Grid Square code e for Grid Square statistics x_e.
2. Convert a Grid Square V_e on the European Reference Grid coordinate system into a Grid Square V_e' on the World Grid Square coordinate system.
3. Generate a set of Grid Square codes $Cover(e)$ for the World Grid Square coordinate system covering a Grid Square V_e'.
4. Generate a set of Grid Squares $W_{w'}$ included in a set of Grid Square codes $Cover(e)$.
5. Compute area ratio $\rho(W_{w'}, V_e')$ from Grid Squares $W_{w'}$ and V_e'.
6. By using Eq. (4.20), compute statistical value y_w for a Grid Square W_w of the World Grid Square coordinate system from a set of area ratios $\rho(W_{w'}, V_e')$ and a set of statistical values x_e for Grid Squares V_e included in a set of Grid Square codes $Cover(w)$.

Figure 4.4 shows Grid Square map representation of GEOSTAT 1 km^2 population grid provided from Eurostat [12] converted into World Grid Square statistics. The depth of color represents the degree of population on each World Grid Square.

4.6 Conclusions

This chapter considered fundamental properties of Grid Square statistics. Grid Square statistics can be expressed as tabular matrix. We confirmed a basic procedure to produce Grid Square statistics from qualitative and quantitative data with geographical information. A contingency table in case of summarization for frequency or categorical data in terms of Grid Square codes becomes Grid Square statistics. In the case of quantitative data, matrix data about representative values in terms of Grid Square codes becomes Grid Square statistics.

A 95% confidence interval of Grid Square statistics statistically inferred from sampled categorical data such as ordinal or nominal scale was derived. We also formalized a problem to adjust an estimation ratio with classification errors. We may evaluate errors of Grid Square statistics statistically inferred from random samples. We mentioned basic arithmetic operations between Grid Square statistics and a method to convert Grid Square statistics produced on one Grid coordinate system into those on other Grid coordinate system. We showed examples to apply conversion procedure from the Australia National Grid coordinate system and the European Reference Grid coordinate system to the World Grid Square coordinate system.

References

1. Grid Square Statistics, Statistics Bureau, Ministry of Internal Affairs and Communications. http://www.stat.go.jp/english/data/mesh/
2. Geospatial statistics, Statistics Bureau, Ministry of Internal Affairs and Communications. http://www.e-stat.go.jp/SG1/estat/toukeiChiri.do?method=init
3. Land Numerical Information download service, Ministry of Land, Infrastructure, Transport, and Tourism. http://nlftp.mlit.go.jp/ksj/index.html
4. Nighttime Lights, NASA Earthdata. https://earthdata.nasa.gov/learn/backgrounders/nighttime-lights
5. JISC, The Japanese National Grid Square codes, JIS X0410:2002
6. Research Institute for World Grid Squares, library. https://www.fttsus.org/worldgrids/en/library/. Accessed 7 Jan 2024
7. Eurostat, GridMaker. https://github.com/eurostat/GridMaker. Accessed 13 Jan 2024
8. National Nested Grid. https://www.anzlic.gov.au/resources/national-nested-grid. Accessed 13 Jan 2024
9. Ordnance Survey. https://www.ordnancesurvey.co.uk/support/the-national-grid.html. Accessed 21 June 2014
10. ISO 19170-1:2021, Geographic information—Discrete Global Grid Systems Specifications—Part 1: Core Reference System and Operations, and Equal Area Earth Reference System. https://www.iso.org/standard/32588.html. Accessed 12 Jan 2024
11. Australian Bureau of Statistics, Australian population grid 2016. https://www.abs.gov.au/AUSSTATS/abs@.nsf/DetailsPage/3218.02016. Accessed 7 Jan 2024
12. Eurostat, GEOSTATS 1 km^2 population grid 2011. https://ec.europa.eu/eurostat/web/gisco/geodata/reference-data/population-distribution-demography/geostat. Accessed 8 Jan 2024

Chapter 5
How to Construct Data Infrastructure Based on Grid Square Statistics

Abstract Grid square statistics—a kind of spatial statistics—is useful as a tool for analyzing, comparing and synthesizing statistics regarding anonymization. This chapter proposes how a data cube (linked open data) of World Grid Square Statistics can be constructed based on data collected from a range of sources, such as official statistics, satellite imagery, and point data from Internet providers. We present a definition of our World Grid Square coding system and a method of generating grid square statistics. We further propose a system for providing grid square statistics in a number of specific areas regarding their context and demonstrate their application to a data visualization and analytics platform.

5.1 Introduction

Location information (geographical information, physical addresses, postal/zip codes, and area polygons) is widely included in business data, such as transactional data, customer data, order data, employment data, machine data, and so on. However, it is not easy to use such business data regarding the location information.

The reasons why such business data is merely used are due to the following technical and social constraints.

1. Too much quantity and variety (Big Data issues)
2. Different data formats and file formats (Syntactic issues for interoperability)
3. Analysis connected with other data is not easy due to different geographic coding systems used in location information (Syntactic issues for interoperability)
4. It cannot be taken outside the orgainization for secondary use since it contains personal information (Policy issues for interoperability; Privacy issue)
5. Heterogeneity of data quality (Data quality issues).

If these unused data can be converted into Grid Square statistics by using a standard method that has connectivity and resuability, it may become a data infrastructure that provide us with distributable, searchable, and calculatable data regarding data quality. Since the amount of data is enormous, the data should be provided as data infrustructure as service. Such a platform can solve various bottlenecks in the fiedls of AI and data unilization.

A.-H. Sato and H. Tsubaki, *Evaluation Platform of Sustainability for Global Systems*,
https://doi.org/10.1007/978-981-97-2296-9_5

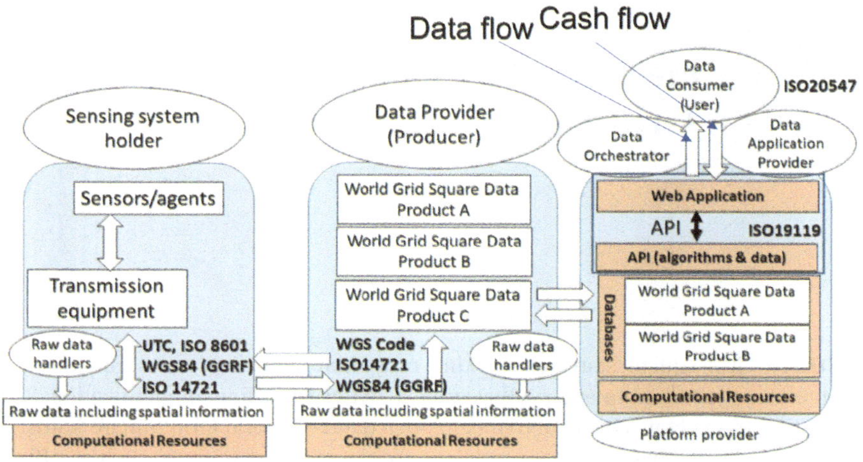

Fig. 5.1 A conceptual illustration of the world grid square data reference architecture

Moreover, it is also difficult to exchange and analyze business data, including the location information related to personally identifiable information (PII), due to easily violating personal data protection laws.

Statistical supression by Grid Square statistics is quite useful in reducing the risk of personal data exchange although Grid Square statistics still connectable through Grid Square codes. In fact, definitions of Grid Square coordination system are various, however, methods to produce, evaluate, and use Grid Square statistics are sharable.

In this chapter, we consider methodological aspects on data infrastructure for Grid Square statistics. We propose World Grid Square Data Reference Architecture (WGSDRA) [1] as shown in Fig. 5.1. The institutional view of WGSDRA consists of four types of organizations:

- Sensing System Holders
- Data Providers (Producer)
- Data Application Providers
- Data Consumers (Users).

Sensing System Holders construct and maintain sensing system to generate original data, including geographic and temporal information. Sensing System Holders need to use the common geodetic datum and time/date expressions to express the data. The context information must be described as expressed as texts. Data Providers aggregate data, including geographic and temporal information, from Sensing System Holders and produce Grid Square statistics. Data Application Providers develop new functions and prepare new types of Grid Square statistics following Data Consumers' demands, and system orchestrators attempt to address the demands of potential Data Consumers and design new functions that should be provided in the Web application.

Table 5.1 A conceptual tabular expression of the services for big data applications

Application type	Use cases
Insect-eye view	(A) simulate the near future, and (B) tell us spatial characteristics
Bird-eye view	(C) visualize what we cannot recognize, and (D) seach for the optimal location

Raw data handlers often access confidential information included in original data. They must observe privacy protection laws in each country. The World Grid Square statistics can be shared with the common Grid Square coding system. The five-layer model shown in Fig. 5.4 is useful for managing various Grid Square statistics in Data Providers.

Numerous Grid Square statistics are stored in database servers in a web application platform, and algorithms implemented as APIs use some Grid Square statistics at the same time to compute indicators and access information. Data Consumers access and use functions implemented in the Web application. The Web application platform can support the decision-making of Data Consumers.

We classify two types of data applications, insect-eye and bird-eye views, based on "scenery" and "holistic", and focus on four types of use cases.

We also discuss distributed platform for Grid Square statistics that can autonomously and decentrally operate a group of data applications.

Table 5.1 shows a conceptual matrix of the service that responds to the needs of the two types of user perspective such as insect-eye and bird-eye views. The purpose of insect-eye view type applications is to provide guidlines for working in the real environment and support in acquiring added value. This type includes applications that (A) simulate the near future based on several possible senaric, and (B) explain spatial characteristics on a focial place. The purpose of bird-eye view is to assist enterprise activities by enabling to see thing that cannot be seen from insect-eye view. This type includes applications that (C) visualize what we cannot recognize by using personal data and global data, and (D) search for the optimal area that meets vague wishes.

5.2 Cost Estimation to Construct Distributed Platform

In this section, we will discuss a cost recovery period and compare three different types of platforms for Grid Square statistics from the business model perspective: The Service Type I is the platform in which all the necessary Grid Square statistics data is purchased in advance. The Service Type II is the platform in which the necessary Grid Square statistics data is locally purchased on demand using APIs. The Service Type III is the platform in which a search problem for multiple regions is solved by using multiple use of Service II. We will estimate costs and cost recovery period for the three type of platforms by regarding their system construction cost and operation cost.

A platform of the Service Type I is constructed by purchasing S kinds of Grid Square statistics in advance at the price P_i ($i = 1, \dots, S$) for Grid Square statistics i. Let us consider a cost recovery period when a data application provider provides insect-eye and bird-eye view type data applications for K_t users at time t, where the number of users changes in time. In this case, the cost recovery period is calculated as follows.

If the sum of the data purchase cost P_i for all the data i and construction cost R is less than sales obtained from the K_t users at unit useage fee c by substracting operation cost m_t for each period. Then cost recovery period T_1^* is given by the lower boundary of T in the following inequality:

$$\sum_{t=1}^{T}(cK_t - m_t) = (c\kappa - M)T > \sum_{i=1}^{S} P_i + R, \tag{5.1}$$

where κ and M are temporal averages of the number of users and operation cost for an operation period $[1, T]$, respectively. From Eq. (5.1), one has

$$T > \frac{\sum_{i=1}^{S} P_i + R}{c\kappa - M}. \tag{5.2}$$

Therefore, we obtain the cost recovery period as

$$T_1^* = \frac{\sum_{i=1}^{S} P_i + R}{c\kappa - M}. \tag{5.3}$$

Next, we consider the Service Type II shown in Fig. 5.2 and estimate its cost recovery period. Suppose that a data application provider operates insect-eye view services with S types of Grid Square statistics data by purchasing the ith data with the cost Q_i ($i = 1, \dots, S$) proportional to data quantity on demand. We assume that the data purchase and operation costs are included in user payment. Therefore, we have the following equation for total cost and income balance,

$$\sum_{t=1}^{T}\left(K_t\left(c - \sum_{i=1}^{S} Q_i\right) - m_t\right) = \left(\kappa\left(c - \sum_{i=1}^{S} Q_i\right) - M\right)T > R, \tag{5.4}$$

where κ and M are temporal averages of the number of users and operation cost for an operational period $[1, T]$, respectively. From Eq. (5.4), we obtain

$$T > \frac{R}{\kappa\left(c - \sum_{i=1}^{S} Q_i\right) - M}, \tag{5.5}$$

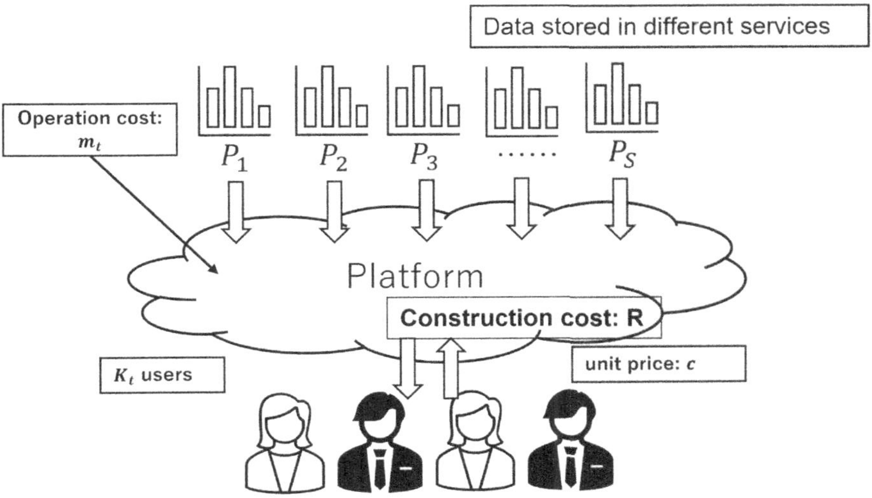

Fig. 5.2 An example of an on-demand data service constructing with several types of grid square data and statistics

and we get the cost recovery period as

$$T_2^* = \frac{R}{\kappa\left(c - \sum_{i=1}^{S} Q_i\right) - M}. \tag{5.6}$$

Finally, let us consider the Service Type III as shown in Fig. 5.3. It means that a bird-eye service can consist of J insect-eye services for different regions. In the case, we estimate a cost of the bird-eye service from J times costs of the insect-eye service. Therefore, we have the following equation for total cost and income balance,

$$\sum_{t=1}^{T}\left(K_t\left(c - J\sum_{i=1}^{S} Q_i\right) - m_t\right) = \kappa T\left(c - J\sum_{i=1}^{S} Q_i\right) - MT > R, \tag{5.7}$$

where κ and M are averages of the number of users and operation cost for an operation period $[1, T]$, respectively. From Eq. (5.4), we obtain

$$T > \frac{R}{\kappa\left(c - J\sum_{i=1}^{S} Q_i\right) - M}, \tag{5.8}$$

we get the cost recover period as

$$T_3^* = \frac{R}{\kappa\left(c - J\sum_{i=1}^{S} Q_i\right) - M}. \tag{5.9}$$

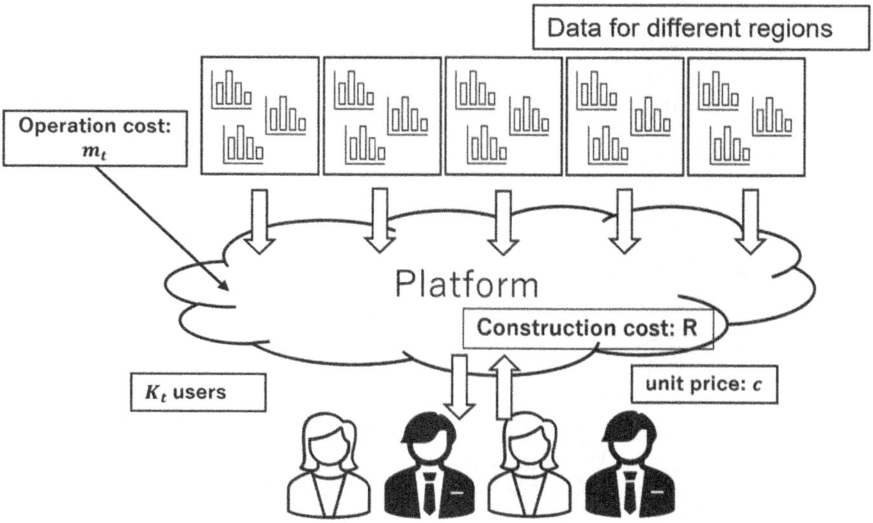

Fig. 5.3 An example of an on-demand bird-eye service constructed from J insect-eye services for different regions

Table 5.2 Estimated cost payback periods for each of the three types of business models

Business models	Cost payback periods
Service Type I	$T_1^* = \frac{\sum_{i=1}^{S} P_i + R}{c\kappa - M}$
Service Type II	$T_2^* = \frac{R}{\kappa(c - \sum_{i=1}^{S} Q_i) - M}$
Service Type III	$T_3^* = \frac{R}{\kappa(c - J\sum_{i=1}^{S} Q_i) - M}$

We summarize the estimated cost recovery periods for Service Types I to III in Table 5.2.

We found that the partial data purchase cost Q_i is small enough for Services type II and type III comparing with the total data purchase cost P_i imposed on Service Type I. In Service Type I, the data application provider requires capital recovery. However, in Service Types II and III, unit usage fee c from each user is locally balanced with data purchase cost $\sum_{i=1}^{S} Q_i$ or $J\sum_{i=1}^{S} Q_i$ at each transaction.

Assume that $\sum_{i=1}^{S} Q_i$ can be charged for each service user, including the usage fee, and that the denominator of the cost payback period can be set to the same value for Service Type I and Service Types II and III. At this time, the numerator with a cost payback period of T_1^* and the numerator with T_2^*, T_3^* is the total data purchase cost $\sum_{i=1}^{S} P_i$.

There is a difference between the cost recovery period of Service Type I T_1^* and the cost recovery period of Service Type II (or III) T_2^* (as well as T_3^*). By using T_2^* (as well as T_3^*), the cost recovery period of Service Type I can be expressed as

$$T_1^* = \frac{\sum_{i=1}^{S} P_i}{c\kappa - M} + T_2^*.$$

The initial total data purchase cost $\frac{sum_{i=1}^{S} P_i}{c\kappa - M}$ is recovered by gross profit $c\kappa - M$ for the period $[1, T]$. From this discussion, we obviously found $T_1^* \gg T_2^*$ and $T_1^* \gg T_3^*$.

In other words, since Service Types II and III business models are realized much faster to recover construction costs than the Service Type I business model, it is concluded that the risk of project implementation of Service Types II and III is smaller than Service type I.

In this way, if data application providers participate in an autonomous and decentralized Grid Square statistics infrastructure and develop and provide data applications under the common environment where they purchase data on demand, they can reduce the risk of recovering the initial investment of data purchase and system construction.

In addition, data application providers can share a user pool registered in the data infrastructure and can possess the fundamental capabilities of the data infrastructure and the timeliness of data procurement.

Through these synergistic effects, data application providers can achieve a high probability of cost recovery by participating in an autonomous and decentralized Grid Square statistics platform, rather than building and operating the system alone and marketing to attract users by themselves.

5.3 Visualization and Analytics Platform for World Grid Square Statistics

In this section, we discuss how to establish a visualization and analytics platform for World Grid Square statistics. We show several examples of data analysis using these statistics and extract use cases of World Grid Square statistics. We implemented a platform, named MESHSTATS, on a cloud service and three types of virtual servers (database server, Web server, and data collection server), which is a test-bed for World Grid Square statistics. MESHSTATS is available from [3].

MESHSTATS provides data and analytics functions. Users can access the service from their own Web browser. This enables users to use provided data and functions implemented in JavaScript under any operating system, which is a reasonable assumption for a multi-operational platform. MESHSTATS adopts super node P2P architecture and data and software migration model for clients. Since MESHSTATS can also provide data through their Web API, users can access the services from their own computer software. Furthermore, automatic data extraction and computation can be realized using the Web API.

As mentioned in Chap. 1, Fig. 5.4 shows a conceptual illustration of the five layers (use cases, data applications, databases and algorithms, a platform, and computational

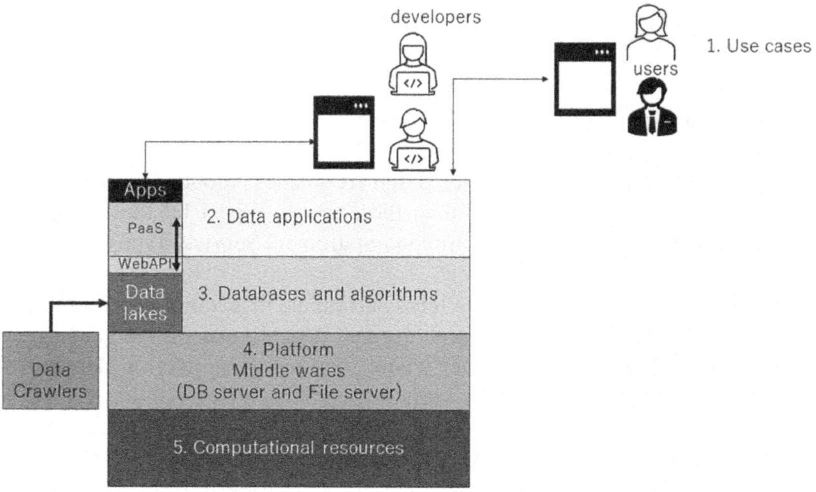

Fig. 5.4 The layered model for the system to integrate geospatial big data. The model consists of five components such as use cases, data applications, databases and algorithms, a platform, and computational resources

resources) for platform for Grid Square statistics. This layer model also follows international standard about Big data reference archtecture in ISO/IEC 20547-3:2020 [2].

Figure 5.5 shows the block diagram of MESHSTATS. Grid Square statistics data is stored separately in a number of databases and segmented into different tables by country. A management unit selects adequate tables of World Grid Squares and extracts the data and statistics as Grid Squares indicated by geographical location (the latitude/longitude information and a focal diameter, World Grid Square codes, physical address, so on). MESHSTATS has two types of output interface such as HTML and API outputs.

In the case of HTML output, the Grid Square statistics extracted by the management unit for Grid Square statistics are converted and displayed as HTML table expression with analytics software. In the case of API output, the Grid Square statistics extracted by the management unit are displayed in a computer-readable format such as CSV.

Web data applications correspond to HTML Unit, HTML I/O, and JavaScript for data analytics. Databases and algorithms consist of API seach query parser, API function, Management Unit for Woeld Grid Squares, and database for World Grid Square statistics. Platform functions are realized by Management Unit for Multi-languages, Language DB, Management Unit for users, and User DB.

The management unit includes a database for users and selects a language based on user settings. A multi-language management unit selects terms according to the user's language setting; field names and expressions on the platform are translated by a dictionary into the user-selected language. The language DB (multi-language

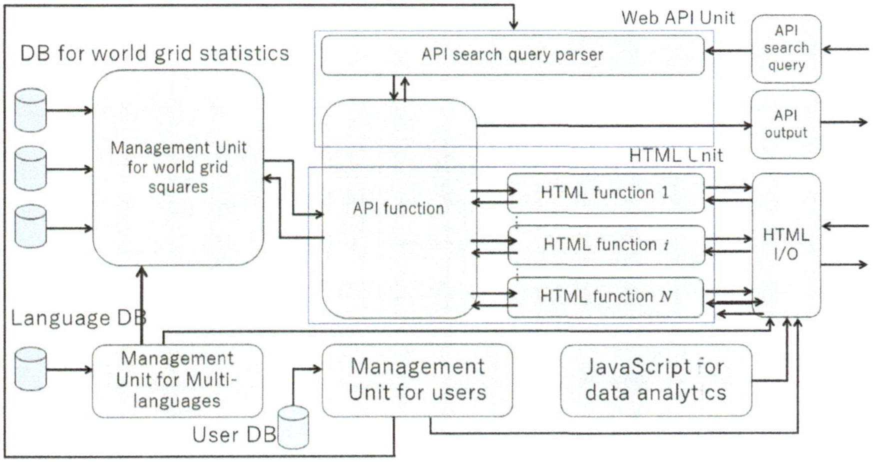

Fig. 5.5 A conceptual illustration of MESHSTATS, which provides functions about the visualization and data analytics for world grid square statistics

dictionary) is constructed by both an API of automatic language translation functions provided by Microsoft Azure and human translators via Web interface. System administrators manually produce texts in English and Japanese and translate them automatically into other languages using the translation API. A human translator who knows English or Japanese but whose mother tongue is other than English or Japanese translates texts into their own language, validating and translating them manually.

At present, our platform supports nine languages (English, Japanese, German, Italian, Spanish, Korean, Vietnamese, Chinese, and Polish). Figure 5.6 shows screenshots of the data visualization and analytics platform. The Grid Square map shows Grid Square statistics on forests, population census, and night-time light intensities in Kyoto, Japan. These screenshots are parts of basic functions provided in MESHSTATS. MESHSTATS further provides various types of Web data applications including data obtained from Web APIs.

5.4 How to Produce World Grid Square Statistics Based on LOD

Since 2016, National Statistics Center of Japan has provided official statistics of Japan based on Linked Open Data (LOD) at an official statistics portal called e-Stat [6].

LOD are presented using standard technologies based on World Wide Web Consortium (W3C) recommendations and are created using the Resource Description Framework (RDF). Normally, LOD are searched using SPARQL Language

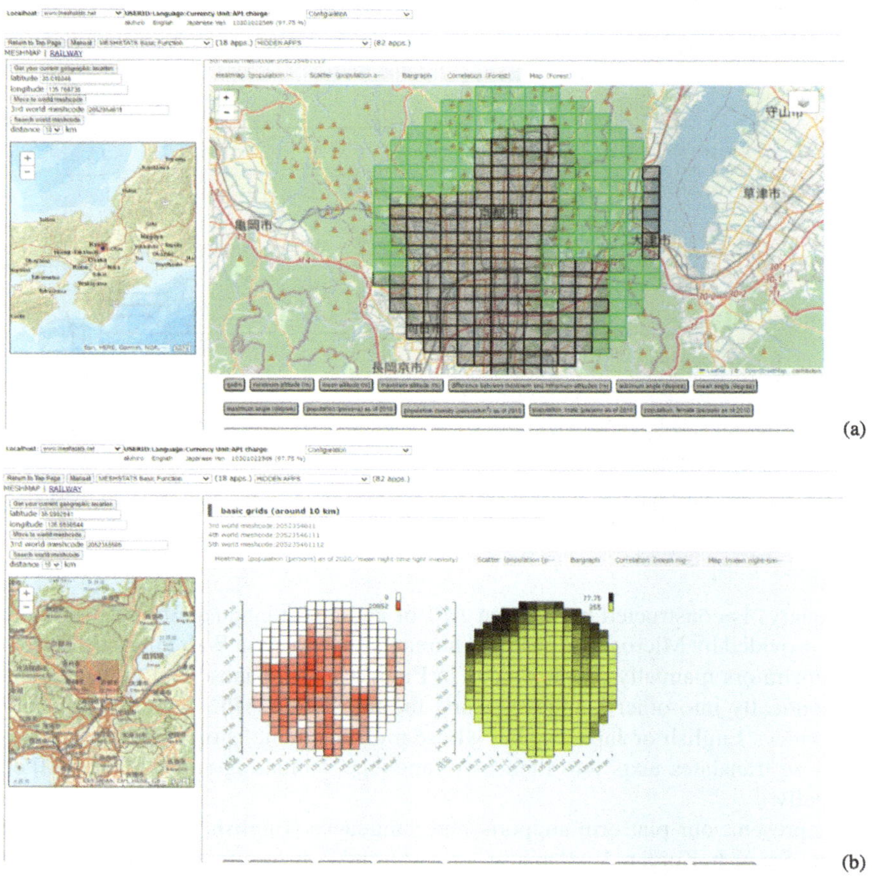

Fig. 5.6 Screenshots of MESHSTATS. The interface provides data visualization and analytics functions. Users can select data types and applications through GUI on Web browser. **a** Grid Square map for green cover ratio. **b** Grid Square maps for population census in 2020 (left) and night-time light intensity in 2018 (right)

(SPARQL). The e-Stat portal has released various official statistics expressed as multi-dimensional data, using RDF formats employing the RDF Data Cube Vocabulary recommended by W3C.

The RDF Data Cube Vocabulary recommends that statistics should comprise four elements: dimension, measure, observation, and attribute. *Dimension* expresses the categories in which statistics are created, such as region, time, gender, and age. *Measure* shows the quantities of statistics produced, such as persons, price, and indices. *Observation* is numerical, expressing the statistics themselves. *Attribute* expresses additional information such as unit and conditions of observation.

The observation in each cell can be expressed by its own URI, and attributes such as units and conditions are linked to the vocabulary databases. The National Statistics Center of Japan has released part of the official statistics (specifically, population census) in RDF format through e-Stat.

Grid Square statistics have not yet been released as LOD. However, the same procedure used for existing trials of official statistics released in RDF formats is applicable in releasing official statistics in Grid Square form. The National Statistics Center plans to release Grid Square statistics that include the population census as LOD. As a Grid Square code is equivalent to a regional unit, the Grid Square code is identified as an entity belonging to the dimension. We can describe a World Grid Square code as a unique URI, and the World Grid Square code can be linked to basic (geographic) information. Data definition can be standardized by providing RDF formatted LOD as URI. Moreover, as a World Grid Square code can be expressed as a unique URI, World Grid Square statistics can readily be linked to data other than official statistics. This enables World Grid Square codes to be used as a primary key to extract data from several databases.

5.5 Probable Data Sources

We identified three types of data source for World Grid Square statistics: official statistics, satellite imagery, and point data collected from the Internet. For example, Global Soil Information Facilities provide gridded data originated from satellite imagery [4], and OpenStreetMap provides point data about objects [5]. These data can be used as primary data to produce Grid Square statistics.

We further found four types of procedure for producing World Grid Square statistics.

1. Convert Grid Square statistics provided as part of official statistics into World Grid Square statistics.
2. Aggregate gridded data and compute World Grid Square statistics.
3. Compute World Grid Square statistics from data that include geographical information (latitude and longitude, World Grid Square codes, and physical address).
4. Generate World Grid Square statistics from polygon data by checking their inclusiveness.

Table 5.3 lists examples of World Grid Square statistics that provided from MESHSTATS. MESHSTATS collects World Grid Square statistics mainly in Japan; statistics and data for other countries were produced from GADM and satellite data sources (land cover classification and night-time light intensity). We may increase the number of types of World Grid Square statistics in countries other than Japan by adding further Grid Square statistics and data to the platform.

Table 5.3 A list of World Grid Square data provided from MESHSTATS

Data	Type	Unit	Data source
Japanese Census Population	Measure	Persons	Population Census, Statistics Bureau of Japan, Ministry of Internal Affairs and Communications [9]
Japanese Census Male Population	Measure	Persons	Population Census, Statistics Bureau of Japan, Ministry of Internal Affairs and Communications
Japanese Census Female Population	Measure	Persons	Population Census, Statistics Bureau of Japan, Ministry of Internal Affairs and Communications
Japanese Workers	Measure	Persons	Economic Census, Statistics Bureau of Japan, Ministry of Internal Affairs and Communications
Japanese Firms	Measure	Firms	Economic Census, Statistics Bureau of Japan, Ministry of Internal Affairs and Communications
Global Administrative Areas	Dimension	–	Global Administrative Area (GADM) [10]
Global Elevation	Measure	Meters	ALOS 30 m elevation, Japan Aerospace Exploration Agency (JAXA) [11]
Japanese Hotels	Measure	Organizations	Accommodation Survey, Japan Tourism Agency, Ministry of Land, Infrastructure, Transport, and Tourism
Japanese Hotel Capacity (Rooms and Beds)	Measure	Rooms	Accommodation Survey of Japan Tourism Agency, Ministry of Land, Infrastructure, Transport, and Tourism
Japanese Overnight Stays	Measure	Persons	Accommodation Survey of Japan Tourism Agency, Ministry of Land, Infrastructure, Transport, and Tourism
Japanese Overnight Stays (Domestic visitors by prefecture)	Measure	Persons	Accommodation Survey of Japan Tourism Agency, Ministry of Land, Infrastructure, Transport, and Tourism
Japanese Overnight Stays (Foreign visitors by countries)	Measure	Persons	Accommodation Survey of Japan Tourism Agency, Ministry of Land, Infrastructure, Transport, and Tourism
Tsunami Hazard	Measure	1/year	NOAA tsunami run-up catalogue data and global elevation by ASTER
Japanese Seismic Hazard	Measure	1/year	Seismic hazard provided by J-SHIS
Japanese Anticipated Flood depth class	Measure	1/year	Ministry of Land, Infrastructure, Transport and Tourism
Global Land cover (20 classes)	Dimension	–	Geospatial Information Authority of Japan (GLCNMO)
Global Night-time light intensity	Measure	–	NASA
Japanese Job advertisement ads by job occupation	Measure	Count	Recruit Web Service
Japanese Room opportunity	Measure	Count	Recruit Web Service

5.6 How to Develop Data Applications

To develop data applications we use a PDCA cycle [7, 8] consisting of data acquisition, data collection, data analysis, interpretation, and deployment.

Section 5.3 considered data visualization and analytics platform from a structural perspective and relationship between platform and roles. Here, we discuss Data Providers, Data Consumers, and Data Application Providers from a constructive point of view. Normally, there is almost always no data at first and the needs are unknown when Data Application Provider develops a data application. As a result, Data Application Providers face the egg-and-chicken problem that they cannot build a data application because they do not have data. They cannot build a data application. Moreover, Data Application Providers do not know the needs of Data Consumers because Data Application Providers do not know who Data Consumers are.

In order to solve this deadlock situation and proceed to the development of a data application, it is necessary for Data Providers to open up their data, Data Application Providers to build data applications using low-value open data, and Data Consumers to use low-value data applications to discover needs. By circulating this work as a PDCA cycle, high-value data is eventually produced, the value of data applications increases, and the number of Data Consumers may increases while advanced needs are discovered.

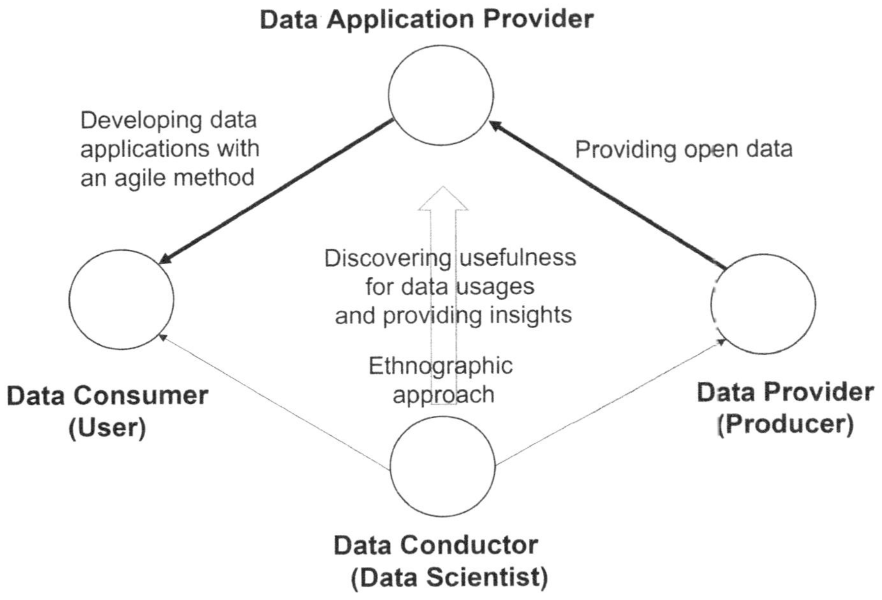

Fig. 5.7 The conceptual illustration of data application development from four types of roles (Data Conductor, Data Provider, Data Consumer, and Data Application Providers)

Figure 5.7 shows a conceptual illustration of data application development from four types of roles. The Data Conductor (Data scientist) needs to be involved in the activities of involving and organizing the entire system with a comprehensive approach even though there is no data supply and no user need. The role of Data Conductor is to find needs eventually by getting into the activities of the people who are Data Consumer and Data Providers, and to discover potentially unused data and deliver it to Data Application Providers eventually. Through the activities of such Data Conductors, Data Application Providers can untangle the deadlock situation where they have no data, do not know the needs of Data Consumers, and cannot find Data Consumer. It is effective for Data Conductors to actually get involved in the activities of potential Data Consumers and Data Providers (those who have data) and to find specifications. This is a kind of agile development, a method that Data Conductors can make possible by their entering the field through an ethnographic approach.

5.7 Conclusion

We discussed how data analytics can employ worldwide spatial statistics based on our recently proposed World Grid Square coding system. We have briefly elaborated a definition of World Grid Square statistics and some examples of their application to administrative areas and elevations. We also addressed use cases for data production based on World Grid Square statistics, referring to some case studies of data applications.

The chapter introduced our data visualization and analytics platform, named MESHSTATS as a cloud-based distributed platform for World Grid Square statistics. MESHSTATS provides visualization, data analytics, and quantification functions for several application domains. In developing functions for the proposed platform, we use a PDCA cycle comprising data acquisition, data collection, data analysis, interpretation, and deployment. We need to develop a function based on domain knowledge and use cases for applications, and to increase the number of Data Providers (Producers) and Data Consumers (Users).

In future work, we will focus on increasing the variety of data providers by developing software to compute Grid Square statistics from primary data and by sharing knowledge to produce World Grid Square statistics. We would emphasize the importance of organizing stakeholders who can provide primary data and/or produce World Grid Square statistics from their own data, create use cases for data applications of the world grid squares, implement functions on the platform, and evaluate work flows, including data analysis of World Grid Square statistics in actual environments.

References

1. A.-H. Sato, S. Nishimura, T. Namiki, N. Makita, H. Tsubaki, World grid square data reference framework and its potential applications, in *Conference: 2018 IEEE 42nd Annual Computer Software and Applications Conference (COMPSAC)* (2018), pp. 398–409. https://doi.org/10.1109/COMPSAC.2018.00062
2. ISO/IEC 20547-3:2020, Information technology—Big data reference architecture. https://www.iso.org/standard/71277.html. Accessed 9 Oct 2023
3. MESHSTATS, Research Institute for World Grid Squares. https://www.meshstats.org/meshstats/. Accessed 7 Jan 2024
4. Global environmental layers. http://worldgrids.org/
5. Open Street Map. http://www.openstreetmap.org/
6. LOD for government statistics, e-Stat, Statistics Bureau in Japan, Ministry of Internal Affairs and Communications. http://data.e-stat.go.jp/lodw/
7. ISO 9001:2015, Quality management systems—Requirements. https://www.iso.org/standard/62085.html. Accessed 15 Jan 2022
8. Ishikawa, K., *What is Total Quality Control? The Japanese Way*. Translated by David J. Lu. (Prentice-Hall, Inc., Englewood Cliffs, NJ, 1985) pp. 56–61
9. Statistics Bureau of Japan, Ministry of Internal Affairs and Communications. https://www.stat.go.jp/english/. Accessed 10 Jan 2024
10. GADM maps and data, GADM. https://gadm.org. Accessed 10 Jan 2024
11. Japan Aerospace Exploration Agency (JAXA). https://global.jaxa.jp/. Accessed 10 Jan 2024

Chapter 6
Distributed Architecture for Grid Square Statistics

Abstract This chapter introduces distributed architecture for storing, searching, and computing Grid Square statistics. Since the data of Grid Square statistics is enormous, we need to consider efficient computation methods under limited computer resources. The distributed architecture gives us many solutions for handling Grid Square statistics. This chapter explains the fundamental principle of parallel computation with MapReduce algorithms or divide-and-conquer methods. We mention Amdahl's law and Gustafson's law. Moreover, we find an implementation of Web API a useful method to exchange data among elements of a distributed system. Finally, we propose a parallelization and distributed architecture for Grid Square statistics.

6.1 Distributed Architecture

A distributed architecture is one of solutions to Big data processing issues. The essential points of the distributed architecture are as follows:

- enlarging data storage
- enlarging processing capacity
- enhancing data throughput
- scalability and elasticity (scale out).

Several types of distributed architectures have been proposed. The typical client-server architecture is classical; multiple client computers communicate with a central server. A peer-to-peer (P2P) architecture forms a decentralized network of peers that play a role with both clients and servers. Several P2P applications such as live video streaming [1], power demand adjustment [2], and blockchain and bitcoins [3, 4] are widely studied.

A.-H. Sato and H. Tsubaki, *Evaluation Platform of Sustainability for Global Systems*,
https://doi.org/10.1007/978-981-97-2296-9_6

6.2 Peer-to-Peer Architecture

There are three models of P2P architecture: pure P2P, hybrid P2P, and centralized P2P. In pure P2P architecture, all the peers (or nodes) are entirely equal. The pure P2P architecture is also called distributed P2P [5]. The hybrid P2P consists of a hypernode network that manages communications among peers that exchange data with other peers. The centralized P2P is that a hyper node manages communications among peers that exchange data with other peers.

Since spatial big data requires sufficient computater resources to handle and process Big data, it is necessary to securely construct the distributed system architecture. Specifically, a centralized P2P architecture must establish a way to confirm other peers in the same P2P network.

Figure 6.1 depicts the centralized P2P system architecture by showing a conceptual illustration of the centralized P2P system through which each peer asks for a common password to the central management server and exchanges data via Web API with the standard password. As shown in Fig. 6.1, Peer E asks peers A and D to obtain data using the Web API of Peer A and Peer D with the standard password. And each peer with a shared password can communicate through Web API.

6.3 Web API

Web API (Application Programming Interface) helps exchange data and functions as Web services. Figure 6.2 shows a conceptual illustration of Web APIs. The API is designed based on the system resources, such as objects, data, and services that users can access. Specifically, the Web API should be indicated as URI uniquely. A Web API is a process having inputs (options) and outputs. The RESTful API over HTTP (HTTPS) is widely used.

Fig. 6.1 A conceptual illustration of the P2P architecture. Programs can access one another via the internet

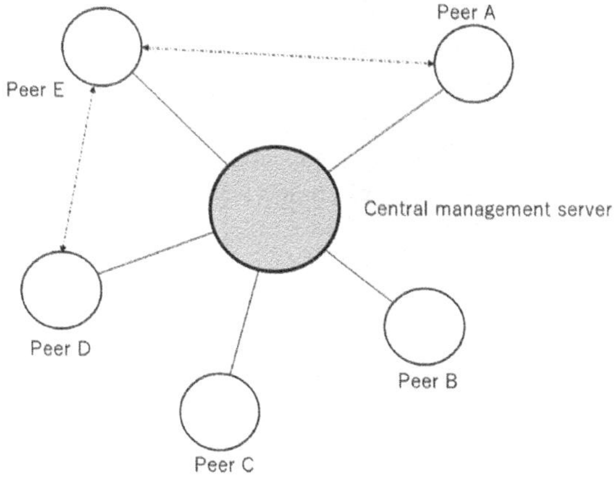

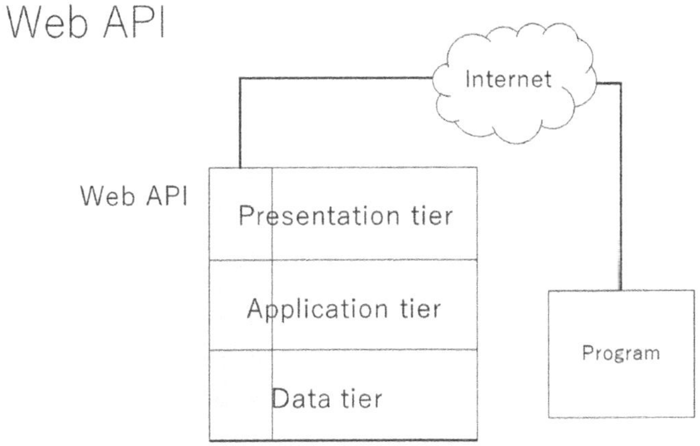

Fig. 6.2 A conceptual illustration of a Web API. A client-side program can access a web API on an application layer via the internet

The REST [11] is an abbreviation of Representation State Transfer, a concept to construct a distributed system based on World Web hypermedia. The Web API is implemented on a Web service (HTTP) application. A web data application is generally based on three layers: data (database), application (logic), and presentation (Web). In this sense, Web APIs provide an interface between Application and Presentation layers, usually implemented over HTTP protocols based on GET and POST.

A client-side program calls a Web API via the Internet and obtains outputs from a Web API. A Web API has several options to provide parameters from the Program. Once the client-side program sends the parameters to the Web API from the Program, the process implemented in the Web API is running at the Web service. Data stored at the Web service is processed based on an algorithm at the Web service and sent outputs to the Program from the Web API via the Internet.

The Web API requires memoryless state changes since we cannot control the order of requests in distributed systems. If the state transition is needed to control, then the outputs may produce the hyperlink to the next state of the Web API, and the Program calls the next state's hyperlink of the Web API.

6.3.1 Options (Inputs) of Web APIs

A client-side program calls a Web API through a URL with some options and parameter values. There are three types of options of Web API (logical, qualitative, and quantitative). An example of URL to access a Web API is as follows:

Table 6.1 An example of options, their type and example value

Options	Type	Example value
`Func`	Qualitative	Default
`Download`	Logical	–
`Latitude`	Quantitative	35
`Longitude`	Quantitative	135
`Distance`	Quantitative	10

https://www.meshstats.xyz/meshstats/api.php?func=defalut&download&latitude=35&longitude=135&distance=10

This example shows an endpoint

https://www.meshstats.xyz/meshstats/api.php

and five options and their value

func=default&download&latitude=35&longitude=135&distance=10,

where the options, their type, and their example value are shown in Table 6.1.

The logical option shows only two states, such as true or false. If the option exists, a function is activated. For example, if option `download` exists, data is downloaded as a CSV file. Otherwise, data is just displayed. The qualitative option means that its parameter is described as letters or codes. This option type often creates a security hole since a code injection attack is applicable. The parameter values of a qualitative option should be sanitized by turning off the string that has the meaning of the command. The most famous attack is an SQL or command injection. Therefore, the qualitative option should be carefully used. The simplest sanitization way is to prepare an acceptable list of strings for parameter values. The quantitative option is described as a numeric sequence. The quantitative option also requires a sanitization process; however, the sanitization algorithm is relatively simpler than the qualitative option. A parameter value of a qualitative option should be cast to integer or float as a sanitization.

6.3.2 Outputs of Web APIs

Outputs of Web APIs are some strings. We need to consider character encoding and a format. Various types of character encodings are defined. A helpful character encoding is UTF-8, which can express multiple letters and characters universally.

The popular output formats of Web API are JSON, XML, TSV, and CSV. JSON is an abbreviation of JavaScript Object Notation, which is used for exchanging data between programs that can express unstructured data such as tree-like representation. An example of JSON format is shown in Fig. 6.2.

Table 6.2 An example of JSON data format

```
{
    "Item name 1": value1,
    "Item name 2": value2,
...
}
```

Table 6.3 An example of XML format data

```
<?xml version'1.0' encoding='UTF-8'?>
<exampledata>
<item ID="001">
<title>item name 1</title>
<value>value1</value>
</item>
<item ID="002">
<title>item name 2</title>
<value>value2</value>
</item>
</exampledata>
```

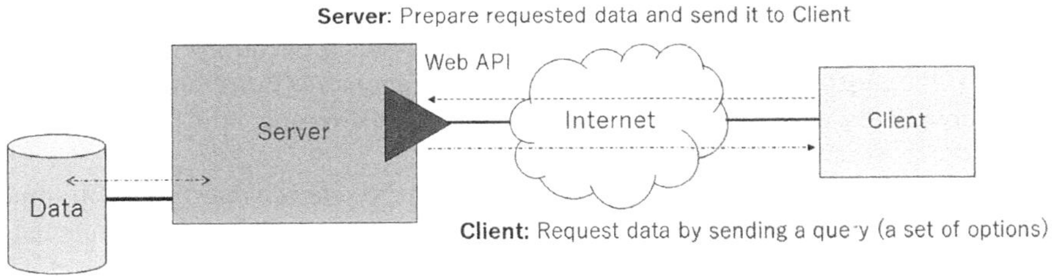

Fig. 6.3 A conceptual illustration of Web API providing data

The XML format also exchanges unstructured data between programs by marking up values with tags with data meanings. Figure 6.3 shows an example of an XML format.

A tab-separated value (TSV) is a format where a tab separates data. The TSV is commonly used to exchange structured data as a table between different application software. A comma-separated value (CSV) is a format where a comma separates data. The CSV is also used to exchange data between application software and the TSV. Both TSV and CSV data require the definition of column name (field name). Table 6.4 shows an example of table data to be expressed in TSV and CSV format.

Table 6.4 An example of table data to be expressed as TSV and CSV format

Worldmeshcode	Filedname1	Fieldname2
Worldmesh1	Value11	Value 12
Worldmesh2	Value21	Value 22
...		

6.4 Example of Web API

This section looks for typical examples of Web API. Grid Square statistics are delivered as structured data formats such as TSV and CSV. Several metadata-related data formats are formatted by unstructured data formats such as JSON and XML. We will find two types of examples for providing Grid Square statistics and automatic original data collection from IoT devices and providing stored data.

6.4.1 Provision of Data

The data provision via Web API is used for many data applications. This example is one of the most familiar cases in the data applications for Grid Square statistics. The Web API has options for selecting data types with several conditions to filter the data.

Calling the Web API with some options provides the selected data type with some conditions. Figure 6.3 shows a conceptual illustration of a Web API to provide data via the Internet.

Table 6.5 shows an example of data provision API.

6.4.2 IoT Devices

IoT devices are used for collecting environmental data with time and location information. IoT devices usually send information of values measured by the sensor to the server via the Internet by calling the server's Web APIs. This API example consists of three APIs, and the data is stored as shown in Fig. 6.4.

These option types are as follows:

- writing data
- selecting data
- searching options.

Tables 6.6, 6.7 and 6.8 show options for three types of API examples.

Table 6.5 A list of options for Web API for providing Grid Square statistics data

Option	Description	Meaning
Date	yyyymmdd	Date is formated as yyyy for year, mm for month, and dd for day
Latitude	Real number ($-90 \leq$ latitude ≤ 90)	Latitude of the central position
Longitude	Real number ($-180 \leq$ latitude ≤ 180)	Longitude of the central positione
Diameter	Real number	Diameter of a circle where data is extracted
Worldmeshcode4	11 decimal digits	4th level of World Grid Square code. If you want to set several World Grid Square codes at the same time, each World Grid Square code are joined by |.
Func	A string	Data type is identified by the func option
Lang_type	Language type	En (English), Ja (Japanese), It (Italian), De (German), Es (Spanish), Ko (Korean), Vi (Vietnamise), Zh (simplified Chinese), Pl (Polish), Th (Thai), Zh-Hant (traditional Chinise), Tr (Turkey), Ar (Arabic), Fr (French), Label (Machine-readable label)

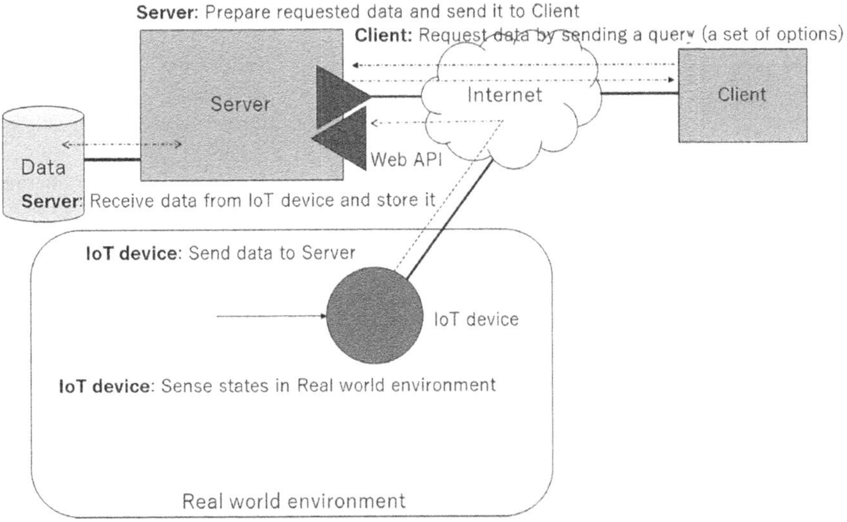

Fig. 6.4 A conceptual illustration of Web API collecting and providing data via IoT

Table 6.6 API for writing data

Option	Description	Meaning
yyyymmdd	yyyy: year, mm: month, dd: day	Date stamp
HHMM	HH: hour, MM: minute	Time stamp
Devid	Device ID	Grid Square code can be used for linking the device with its location
Key	Key string	Key string can be used for identifying value property such as temperature, light intensity, CO_2 concentration and so on
Value	Value string	Value string is data value

Table 6.7 API for selecting data

Option	Description	Meaning
yyyymmdd	yyyy: year, mm: month, dd: day	Available device ID are retrieved for selected date stamp (yyyymmdd)
Devid	Device ID	Available date stamps are retrived for selected device ID (devid)

Table 6.8 API for searching options

Option	Description	Meaning
yyyymmdd	yyyy: year, mm: month, dd: day	Date stamp
HHMM	HH: hour, MM: minute	Time stamp
Devid	Device ID	Grid Square code can be used for linking the device with its location
Key	Key string for selecting data	Key string is used for filtering data by key. If key is set as *, all the records are extracted

As shown in Table 6.6, the API for writing data is called by an IoT device via the Internet with some options to determine timestamp, data key, value, and device ID. As shown in Table 6.7, the API stores the data collected at the sensor device in the server. Users or web application programs may call the API to select data with some options to determine the key string, device ID, and timestamp. The data is used to visualize and analyze the status of sensing devices and the environment. As shown in Table 6.8, the API for searching options is used for retrieving acceptable options from a list of data stored in the server. The API for searching options is used to obtain a master list of device IDs and timestamp options.

6.5 Effective Parallel Algorithms

The distributed system enables the storage of data in a separate manner and the application of parallel computation algorithms. Divide-and-conquer method for parallelization [12] or MapReduce algorithms [13, 14] are effective if Gustafson's law is satisfied [18].

The Gustafson's law realizes a superlinearity between the number of processors and computation duration. That is, when a large amount of computer resources is available for processing and calculating a large amount of data in parallel, it is possible to increase the parallelization coefficient linearly with the number of processors almost in proportion.

The Divide-and-conquer method for parallelization or MapReduce algorithm applies to various types of linear computation. For example, under a parallelization with N cores, summation S for quantitative data $\{x_1, \ldots, x_M\}$ with M sample size can be expressed as

$$S = \sum_{i=1}^{M} x_i = \sum_{k=1}^{N} \left(\sum_{i=m_{k-1}+1}^{m_k} x_i \right), \tag{6.1}$$

where $0 = m_0 < m_1 < \cdots < m_N = M$.

More generally, it is formalized as follows: Suppose that $\{x_1, \ldots, x_M\}$ are data with sample size M. If we can express process function $f(x_1, \ldots, x_M)$ with aggregation function $g(y_0, \ldots, y_{N-1})$ as

$$f(x_1, \ldots, x_M) = g\Big(f(x_1, \ldots, x_{m_1}), \ldots, f(x_{m_1+1}, \ldots, x_{m_2}), \ldots, f(x_{m_{N-1}+1}, \ldots, x_M)\Big), \tag{6.2}$$

then we can compute the process function in a parallel manner. Specifically, the process function f is called mapper or divide, and aggregation function g is called reducer or conquer in the MapReduce algorithm or divide-and-conquer method. The computational algorithms applicable to the MapReduce algorithm and divide-and-conquer method should be, in general, linear calculations such as summation, mean, variance, minimum, maximum, and searching values in a certain range, and extract records from the condition, also satisfies Eq. (6.2). Applicable algorithms are widely recognized, such as the merge-sort method, prefix sum (or scan), matrix multiplication [15], matrix inversion [16], and Apriori algorithm for a association rule-mining problem [17].

In parallel computation, speed-up is one of the most popular performance indicators to measure the efficiency of parallelization, which is given by the ratio of time taken by a single processor system and that taken by a parallel processing system.

The Speed-up is defined as

$$\text{Speed-up} = \frac{T_1}{T_N}, \tag{6.3}$$

where N is the number of processors, T_1 and T_N are the single processor execution time and the N processor execution time, respectively.

According to Amdahl's law [20], it is assumed that the ratio of processing that cannot be parallelized r_s and the ratio of processing that can be parallelized r_p are independent of the number of processors N used for parallel processing, where $r_s + r_p = 1$. The According to Amdahl's law, expressed as

$$\text{Speed-up} = \frac{1}{r_s + r_p/N} = \frac{1}{r_s + (1 - r_s)/N}, \tag{6.4}$$

the coefficient of parallelization saturates when the number of processors N used for parallel processing is increased beyond a certain number.

In several types of computations, parallelization can not only improve efficiency by parallelization, but also reduce the time required for processing on each processor by dividing data exclusively to each processor during parallelization or limiting the search data area without duplication to each processor and assigning processing in a parallel distributed manner by reducing the search area.

The parallelization coefficient in Gustafson's Law [19] is expressed as

$$\text{Speed-up} = r_s + Nr_p = N - r_s(N - 1) \tag{6.5}$$

Equation 6.5 shows a superlinearity of parallelization coefficient regarding the number of processors N. The derivation of Gustafson's law is as follows. Suppose that the parallel execution time with N processors is expressed as

$$T_N = W(r_s + r_p), \tag{6.6}$$

where W is a unit workload. r_s and r_p are the friction where the algorithm cannot be parallelized and can be parallelized, respectively ($r_s + r_p = 1$). If the single processors conduct the same computation with a single process, then single execution time Wr_s requires the same execution time. However, the parallel execution time Wr_p requires N times larger than in the parallel computation. Thus, we have

$$T_1 = W(r_s + Nr_p). \tag{6.7}$$

By using Eqs. (6.6) and (6.7), we have the same equation as shown in Eq. (6.5):

$$\text{Speed-up} = \frac{W(r_s + Nr_p)}{W(r_s + r_p)} = r_s + N(1 - r_s) = N - r_s(N - 1). \tag{6.8}$$

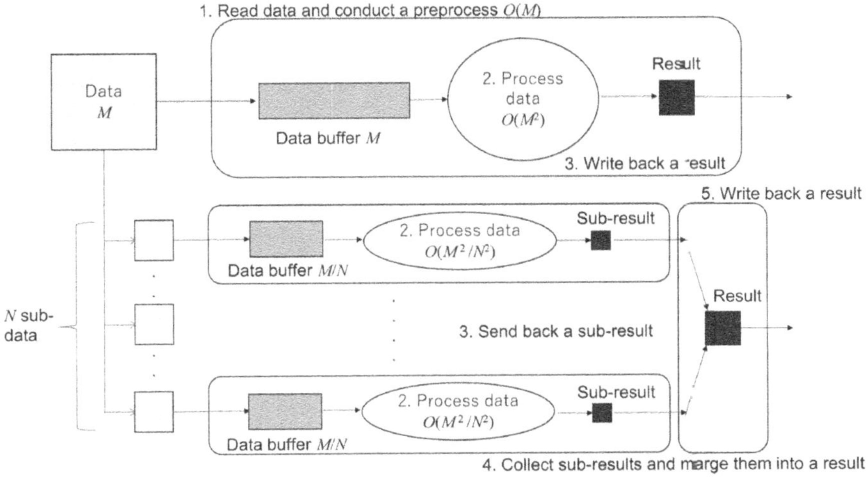

Fig. 6.5 A conceptual illustration comparing between the single data processing and a parallel data processing

We attempt to derive the speed-up coefficient under a more realistic assumption than Eq. (6.5). Suppose that a large amount of data stored in a large file is divided into N equal parts in advance and passed to N processors. Let M be the total amount of the dataset. N processors read and process $1/N$ times the size of the data and integrated to form the final result. In this case, an equation similar to Gustafson's law shown in Eq.(6.5) is derived.

Consider dividing the data with M volume into N parts and processing it by delivering it to N processors. Let W be the task load and T be the processing time per unit task load at a processor. Figure 6.5 shows a conceptual illustration comparing single and parallel data processing. A single processor processes the data with volume M under $O(M^2)$ algorithm and writes back a result in the above case shown in Fig. 6.5. N processors process data divided into N parts under $O(M^2/N^2)$ algorithm and results are merged and written back in the bottom case shown in Fig. 6.5.

We assume that the processing algorithm that is not parallelized is $O(M)$ concerning the data amount M, and the processing algorithm to be parallelized is $O(M^2)$ concerning the data amount M. We estimate the execution time T_1 that cannot be computed in parallel and the execution time T_N that can be calculated in parallel in terms of N and M.

Usually, the execution time to read data from a file is treated as a part of the process that is not computed in parallel. However, using a distributed or distributed database system dividing the data file into N parts in advance, the execution time to read data and send them to each processor in parallel becomes a tiny value. Then, $O(M)$ data preprocessing, such as reading data from a file, which is the majority of processing that cannot be parallelized, can be parallelized. In this way, when the amount of data M/N is assigned to each processor with a small transfer time c_s every processor processes the assigned M/N data in parallel. We assume that the execution time in parallel is proportional to both $O(M/N)$ and $O(M^2/N^2)$. Therefore, we may approximate that the execution time in parallel and the execution time in a single processor is given by $T_N = c_s + T_1 r_p/N + T_1 r_p/N^2$ and T_1, respectively. The parallel coefficient is calculated as

$$
\begin{aligned}
\text{Speed-up} &= \frac{T_1}{c_s + T_1 \frac{r_p}{N} + T_1 \frac{r_p}{N^2}} \\
&= \frac{T_1}{c_s + T_1 \frac{r_p}{N} + T_1 \frac{r_p}{N^2}} \\
&= \frac{1}{c_s/(NT_1) + \frac{r_p}{N} + \frac{r_p}{N^2}} \\
&= \frac{N}{c_s/(NT_1) + r_p + \frac{r_p}{N}} \\
&\approx \frac{N/r_p}{1 + \frac{1}{N}} \\
&= \frac{N}{r_p}(1 - \frac{1}{N}) \\
&= \frac{N}{r_p} - \frac{1}{r_p} \\
&= \frac{N}{1 - r_s} - \frac{1}{1 - r_s},
\end{aligned}
$$

where we assume that $c_s/(NT_1) \approx 0$ and may approximate for $r_p N \gg 1$,

$$
\frac{1}{1 + \frac{1}{r_p} \cdot \frac{1}{N}} \approx 1 - \frac{1}{r_p} \cdot \frac{1}{N}. \tag{6.9}
$$

Equation (6.9) is also rewritten as

$$
\text{Speed-up} = \alpha N - \alpha, \tag{6.10}
$$

where we set $\alpha = 1/r_p = 1/(1 - r_s)$.

6.6 Existing Distributed Database Architectures

DynamoDB [6, 7] by Amazon, the Google File System (GFS) [8] by Google, Apache Hadoop [9], and the Cassandra [10] by Facebook are the most successful distributed databases. DynamoDB introduces a membership algorithm that helps every node maintain information about every other node. The GFS uses a simple design with a single primary server for hosting all the metadata, and the data is split into chunks and stored in chunk-distributed servers. The Apache Hadoop provides a framework for the distributed processing of large datasets across computer clusters based on simple programming models. The Cassandra can handle a table-distributed multi-dimensional map indexed by a key. Every operation is done by a single key expressed by a string in a table. It implies that data can only be handled by reading the contents of a few columns. The Cassandra API consists of three simple methods:

- *insert(table, key, rowMutation)*
- *read(table, key, columunName)*
- *delete(table, key, columnName).*

where *key* identifies the data item and is assigned to a node by hashing the data item's key. Moreover, Cassandra uses the following distributed systems techniques:

- partitioning
- replication
- membership
- failure handling
- scaling.

Partitioning means the ability to dynamically partition the data over the set of nodes in the cluster. Replication is the ability to replicate each data item into N hosts. Cluster membership is determined based on an anti-entropy mechanism and used for failure handling. Scaling means automatically adjusting the number of nodes in the cluster.

6.7 Distributed Architecture for Grid Square Statistics

We propose a distributed architecture for Gird Square statistics. As mentioned in Sect. 6.5, effective parallel algorithms are based on Eq. (6.2). The fundamental concept of our architecture is that a unique Grid Square code is used for a key in distributed databases like Key-Value distributed databases.

Figure 6.6 shows a conceptual illustration of the distributed database for Grid Square statistics. Each node contains hierarchical directory trees consisting of table names and the first level of Grid Square codes. Each directory named as the first level of Grid Square code contains files including table data about Grid Square statistics in each level of Grid Square. The distributed database for Grid Square statistics has three API methods, like Key-Value distributed databases.

- *insert(table, GridSquareCode, rowMutation)*
- *read(table, A set of GridSquareCodes, columunName)*
- *delete(table, GridSquareCode, columnNmae).*

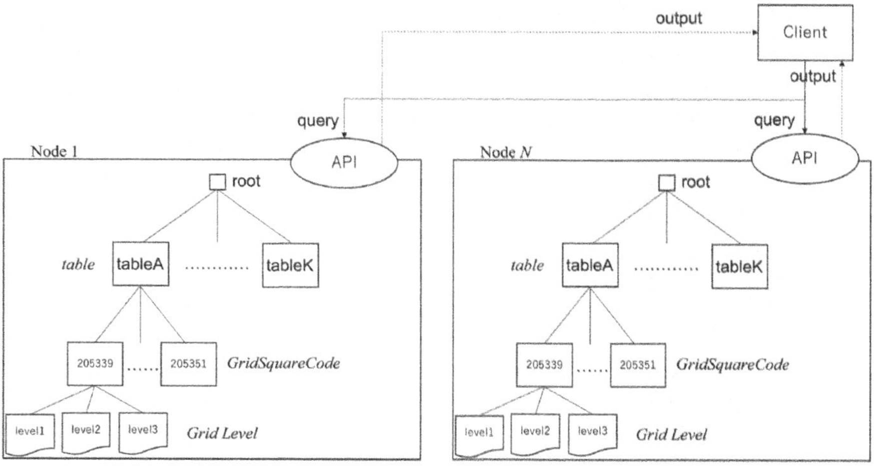

Fig. 6.6 A conceptual illustration of the distributed database for Grid Square statistics. A client sends a query to each API and obtain a result by merging outputs searched in files stored in hierarchical directories by each API

As shown in Fig. 6.6, the searching algorithm is as follows:

- A client sends the same query to each API.
- Each API searches files named the first level of Grid Square codes computed from *GridSquareCode*(s) in *table*.
- Each API extracts records with *GridSquareCode*(s) and output a value in *columnName* with *GridSquareCode*.
- A client merges values with *GridSquareCode* as record.

6.8 Conclusion

This chapter explained the distributed architecture and fundamental principles of efficient parallel computation for Grid Square statistics. The distributed architecture gives us many solutions to solve computational issues we face when using Grid Square statistics.

Web API is a solution for parallel computations under Web-based distributed architecture. We showed an example of Web API and types of Web API options. Web API can connect internal functions in Web servers with external functions in external programs (client-side Program in Web browser and user outer program). We discussed two types of Web API scenarios, such as the provision of data and IoT devices.

If we limit algorithms for parallel computation as MapReduce algorithms, we may realize superlinearity between computational resources and computation time. Since Grid Square statistics require massive computational resources to produce, search, and calculate them, we need computational resources to store, search, and compute them efficiently. We derived an approximated equation of parallel coefficient (speed-up) for Gustafson's scenario. If we use the MapReduce algorithm or the divide-and-conquer method for parallelization for data processing, the speed-up is proportional to the number of parallel components under the condition of a sufficient amount of data.

This chapter proposed a distributed database for Grid Square statistics via Web API. Our proposed system architecture for Grid Square statistics is the system architecture where a Grid Square code is used as a primary key for Web-based key-value distributed databases. The distributed architecture gives us many solutions to solve computational issues that we face when we use Grid Square statistics.

References

1. Y. Liu, Y. Guo, C. Liang, A survey on peer-to-peer video streaming systems. Peer-to-Peer Netw. Appl. **1**, 18–28 (2008)
2. E. Pournaras, M. Warnier, F.M.T. Brazier, Peer-to-peer aggregation for dynamic adjustments in power demand. Peer-to-Peer Netw. Appl. **8**, 189–202 (2015)
3. S. Nakamoto, Bitcoin: A peer-to-peer electronic cash system. Decentralized Bus. Rev. 21260 (2008)
4. J. Bonneau, A. Miller, J. Clark, A. Narayanan, J.A. Kroll, E.W. Felten, Sok: Research perspectives and challenges for bitcoin and cryptocurrencies, in *IEEE Symposium on Security and Privacy*. (IEEE, 2015), pp. 104–121
5. J. Zhang, The architecture of P2P computer collaborative design system based on artificial intelligence. J. Phys.: Conf. Ser. ICCASIT 2020 **1574**, 012090 (2020). https://doi.org/10.1088/1742-6596/1574/1/012090
6. G. DeCandia, D. Hastorun, M. Jampani, G. Kakulapati, A. Lakshman, A. Pilchin, S. Sivasubramanian, P. Vosshall, W. Vogels, Dynamo: amazon's highly available key-value store. ACM SIGOPS, Oper. Syst. Rev. 41(6), pp. 205–220 (2007). https://doi.org/10.1145/1323293.1294281
7. AWS, Amazon DynamoDB. https://docs.aws.amazon.com/amazondynamodb/latest/developerguide/Introduction.html. Accessed 25 Dec. 2023
8. S. Ghemawat, H. Gobioff, S.-T. Leung, The google file system, in *SOSP '03: Proceedings of the Nineteenth ACM Symposium on Operating Systems Principles* (2003), pp. 29–43. https://dl.acm.org/doi/10.1145/945445.945450
9. Apache Hadoop. https://hadoop.apache.org. Accessed 25 Dec 2023
10. A. Lakshman, P. Malik, Cassandra: a decentralized structured storage system. Oper. Syst. Rev. **44**(2), 35–40 (2010)
11. RESTful web API design, Microsoft, Azure, Product Documentation. https://learn.microsoft.com/ja-jp/azure/architecture/best-practices/api-design
12. V. Niculescu, On generalizing divide and conquer parallel programming pattern. Mathematics **10**(21), 3929 (2022). https://doi.org/10.3390/meth10213925
13. J. Dean, S. Ghemawat, MapReduce: simplified data processing on large clusters. *OSDI'04: Sixth Symposium on Operating System Design and Implementation* (CA, San Francisco, 2004), pp. 137–150

14. S. Maitrey, C.K. Jha, MapReduce: simplified data analysis of big data. Proceedia Comput. Sci. **57**, 563–571 (2015)
15. R. Bagheri, M. Amini, S.D. Samarin, Efficient verification of parallel matrix multiplication in public cloud: the MapReduce case. J. Big Data **7**, 86 (2020). https://doi.org/10.1186/s40537-020-00362-1
16. J. Xiang, H. Meng, A. Aboulnaga, Scalable matrix inversion using mapreduce, in *HPDC '14: Proceedings of the 23rd International Symposium on High-performance Parallel and Distributed computing* (2014), pp. 177–190. https://doi.org/10.1145/2600212.2600220
17. P. Kijsanayothin, G. Chalumporn, R. Hewett, On using mapreduce to scale algorithms for Big Data analytics: a case study. J. Big Data **6**, 105 (2019). https://doi.org/10.1186/s40537-019-0269-1
18. R.E. Benner, J.L. Gustafson, G.R. Montry, Development and analysis of scientific application programs on a 1024-processor hypercube, SAND 88-0317 (Sandia National Laboratories, 1988)
19. J.L. Gustafson, Gustafson's law, in *Encyclopedia of Parallel Computing*, ed. by D. Padua (Springer, Boston, MA, 2011). https://doi.org/10.1007/978-0-387-09766-4_78
20. G.M. Amdahl, Validity of the single processor approach to achieving large-scale computing capabilities, in *AFIPS Conference Proceedings*, vol. 30 (1967), pp. 483–485

Chapter 7
Producing Grid Square Statistics on Demand

Abstract This chapter proposes an on-demand service to produce Grid Square statistics. This on-demand service is based on the Generic Statistical Business Process Model (GSBPM) developed by UNECE. GSBPM consists of eight phases: Specify Needs, Design, Build, Collect, Process, Analyze, Disseminate, and Evaluate. We prototype an on-demand service for World Grid Square statistics as a web data application based on data collected from the OpenStreetMap (OSM) service. The system conducts processes automatically, from request acceptance to disseminating the World Grid Square statistics. We propose the following four steps to form the on-demand service in real-time: (i) users check out points of interest; (ii) the service automatically accesses Web API of Web Feature Service (WFS) to collect the data on points of interest; (iii) the system computes Grid Square statistics, and (iv) disseminates them. Since the types of points of interest dealt with in WFS vary enormously, and the total number of the World Grid Squares is vast, it is impossible to produce World Grid Square statistics for all the possible types of interests beforehand. However, our proposed service can produce and serve the Grid Square statistics in response to users' requests. Moreover, we discuss a potential extension of the proposed system to other types of geographical expressions, such as postal codes and addresses.

7.1 Introduction

It is important to construct data infrastructure to promote data-driven organizations and society. Specifically, this chapter considers a global data platform to evaluate and visualize sustainability from a global perspective. Such a global data platform requires data collection on socioeconomic activities from a global perspective and its spread as a distributed system to support decision-making.

The global data platform needs to handle Big Data [1], which can be characterized by 4V (Volume, Velocity, Variety, and Variability).

A.-H. Sato and H. Tsubaki, *Evaluation Platform of Sustainability for Global Systems*,
https://doi.org/10.1007/978-981-97-2296-9_7

Big Data is a relative concept between the system's data and information processing capacity. If the data exceeds the information processing capacity available, we need a strategy to manage data scalability. Geospatial data is Big Data and requires a solution to handle such issues. Grid Square statistics can be used to reduce the volume and velocity of original geospatial data.

This chapter discusses an automatic production of Grid Square statistics on demand, which can produce World Grid statistics automatically from data stored in the system driven by users' data search behavior.

First of all, we introduce existing frameworks related to Grid Square statistics. There are several existing frameworks related to data and statistics. Generic Statistical Business Process Model (GSBPM) [2] is developed by UNECE, which consists of eight phases: Specify Needs, Design, Build, Collect, Process, Analyze, Disseminate, and Evaluate.

The United Nations General Assembly adopted a global geodetic reference frame (GGRF) [3, 4] for sustainable development in 2015. The GGRF framework allows users to precisely determine and express locations on the Earth based on common methods and also enables users to quantify changes in the Earth in space and time.

As mentioned in Chap. 1, the Global Statistical Geospatial Framework (GSGF) [5] is adopted in the United Nations, Committee of Experts on Global Geospatial Information Management (UN-GGIM), and the United Nations Statistical Commission (UNSC) and has five fundamental principles:

- P1: Use of fundamental geospatial infrastructure and geocoding
- P2: Geocoded unit record data in a data management environment
- P3: Common geographies for dissemination of statistics
- P4: Interoperable data and metadata standards
- P5: Accessible and usable.

The latitude-longitude information is defined as the World Geodesic System (WGS) and used in cartography, geodesy, and satellite navigation. This forms fundamental geospatial infrastructure and geocoding.

Many country use postal addresses, postal codes (postal code, PIN, or ZIP code) as a common geocoded units. However, they often vary over time. In the case of Japan, standard Grid Squares are defined as artificial common geocoded units based on the geographical information (latitude and longitude). Japan has Japanese National Grid Square coding system defined in JIS X0410:2002 [6].

For example, the basic Japanese National Grid Square codes (the third level) are computed from latitude and longitude from the following equations:

$$\begin{cases} \lfloor \text{latitude} \times 60 \div 40 \rfloor = p \quad (p \text{ is two digits.}), \\ a := (\text{latitude} \times 60 \div 40 - p) \times 40, \\ q := \lfloor a \div 5 \rfloor \quad (q \text{ is one digit.}), \\ b := (a \div 5 - q) \times 5, \\ r := \lfloor b \times 60 \div 30 \rfloor \quad (r \text{ is one digit.}), \\ c := (b \times 60 \div 30 - r) \times 30, \\ \lfloor \text{longitude} - 100 \rfloor = u \quad (u \text{ is two digits.}), \\ f := \text{longitude} - 100 - u, \\ v := \lfloor f \times 60 \div 7.5 \rfloor \quad (v \text{ is one digit.}), \\ g := (f \times 60 \div 7.5 - v) \times 7.5, \\ w := \lfloor g \times 60 \div 45 \rfloor \quad (w \text{ is one digit.}), \\ h := (g \times 60 \div 45 - w) \times 45, \end{cases} \tag{7.1}$$

Finally, the third level of the Japanese National Grid Square code is defined as

$$(\text{Grid Square code}) = puqvrw. \tag{7.2}$$

The JIS X0410:2002 has a hierarchical structure consisting of 6 levels. Administrative areas are used as geographical units in statistical aggregation.

As shown in Chap. 3, the World Grid Square code [7] is an extension compatible with JIS X0410:2002. This is a key definition for generating Grid Square statistics data from a global perspective. Libraries to compute a World Grid Square code from latitude and longitude are provided by five types of computer languages (R, JavaScript, PHP, Python, and Java) [8].

For example, the third level World Grid Square code for a Grid Square including a point indicated by latitude and longitude is defined as follows:

$$o = 2^2x + 2y + z + 1. \tag{7.3}$$

$$z = (o - 1) \bmod 2, \tag{7.4}$$

$$y = ((o - z - 1) \div 2) \bmod 2, \tag{7.5}$$

$$x = (o - 2y - z - 1) \div 4, \tag{7.6}$$

$$\text{Grid Square code} = \begin{cases} o00p0uqvrw & (p < 10, u < 10) \\ o0p0uqvrw & (10 \leq p < 100, u < 10) \\ op0uqvrw & (p \geq 100, u < 10) \\ o00puqvrw & (p < 10, u \geq 10) \\ o0puqvrw & (10 \leq p < 100, u \geq 10) \\ opuqvrw & (p \geq 100, u \geq 10) \end{cases}, \tag{7.7}$$

$$\begin{cases} p := \lfloor (1-2x)\text{latitude} \times 60 \div 40 \rfloor \quad (p \text{ is two or three digits}), \\ a := \left\{(1-2x)\text{latitude} \times 60 \div 40 - p\right\} \times 40, \\ q := \lfloor a \div 5 \rfloor \quad (q \text{ is one digit}), \\ b := (a \div 5 - q) \times 5, \\ r := \lfloor b \times 60 \div 30 \rfloor \quad (r \text{ is one digit}), \\ u := \lfloor (1-2y)\text{longitude} - 100z \rfloor \quad (u \text{ is one or two digits}), \\ f := (1-2y)\text{longitude} - 100z - u, \\ v := \lfloor f \times 60 \div 7.5 \rfloor \quad (v \text{ is one digit}), \\ g := (f \times 60 \div 7.5 - v) \times 7.5, \\ w := \lfloor g \times 60 \div 45 \rfloor \quad (w \text{ is one digit}), \end{cases} \quad . \tag{7.8}$$

where o is called the zero level of the World Grid Square code. p, q, r, u, v, and w are integers computed from Eqs. (7.8). They form the third level of the World Grid Square code, which is expressed as a digit decimal sequence with 30 arc seconds for latitude and 45 arc seconds for longitude.

The World Grid Square is used as an aggregation unit for computing geospatial statistics from spatial big data and as a super key to effectively identifying data groups to search spatial big data (point and raster data).

7.2 Grid Square Statistics

The Grid Square statistics are computed using Grid Squares as an aggregation unit. The number of elements for qualitative data and representative statistics, such as minimum, maximum, median, and mean of quantities for quantitative data, are selected as statistics (see Sect. 4.2).

If we have point data about features, which include a location expressed by latitude and longitude, we can compute Grid Square statistics about the number of the features. We identified three types of data sources: official statistics, satellite imagery, and point data collected from the Internet. As shown in Chap. 4, we found four types of procedures to create World Grid Square statistics:

- Convert Grid Square statistics provided as part of government statistics into Grid Square statistics.
- Aggregate gridded data such as satellite images and compute Grid Square statistics.
- Compute Grid Square statistics from data that include geographical information (latitude/longitude information and postal address).
- Generate Grid Square statistics from polygon data by checking their inclusiveness.

Typically, point data, including detailed locations, are sensitive; we need careful treatment in exchanging and disseminating such point data. We proposed World Grid Square statistics and data reference architecture [9] consisting of three types of organizations: sensing system holders, data providers, and data application providers.

As mentioned in Chap. 5, we propose World Grid Square Data Reference Architecture (WGSDRA). We may consider the structure of an on-demand service on World

Grid Square statistics based on WGSDRA. Sensing system holders generate original data, including geographic and temporal information. Sensing system holders must use the common geodetic datum and time/date expressions to express the data. The context information must be described as expressed as texts. Data providers aggregate data obtained from sensing system holders, including geographic and temporal information, and produce World Grid Square statistics. Data application providers develop new functions and prepare new World Grid Square statistics types following data consumers' demands. System orchestrators attempt to address the demands of potential data consumers and design new functions that should be provided in the web application.

Raw data handlers often access confidential information included in original data. They must observe privacy protection laws in each country. The World Grid Square statistics share the standard World Grid Square coding system under WGS84. The OAIS is useful for managing various World Grid Square statistics in data providers. Numerous World Grid Square statistics are stored in database servers in a web application platform, and algorithms implemented as APIs use some World Grid Square statistics simultaneously to compute indicators and access information. Data consumers access and use functions implemented in the web application.

7.3 Proposed System Architecture

Big point data is a challenge for data analytics. Big point data are bigger than common information processing capacity in the sense of 4V (Volume, Velocity, Variety, and Variability). This implies that some data are possible and producing their World Grid Squares beforehand is impossible. Our proposed system identifies the system requirements to respond to the challenge.

Geospatial information collected by users is automatically stored in the system. Users' interest in Point-of-Interest (POI) can be detected by users collecting behavior. The system produces World Gird Square statistics from the stored data. The users can analyze world Grid Square statistics automatically created from the data collection driven by interest.

Figure 7.1 shows the activity diagram for the proposed system. The system activity has three layers: POI visualization on a map, automatic production of World Grid Square statistics, and visualization and analysis of World Grid Square statistics.

The POI visualization on a map consists of five steps:

1. Select an area and a type of POI.
2. Generate a query.
3. Send a query to get POI to Web Feature Service.
4. Collect POI from Web Feature Service.
5. Visualize POI on a map.

During step 4, POI is stored into a system data lake. The automatic production of World Grid Square statistics consists of three steps:

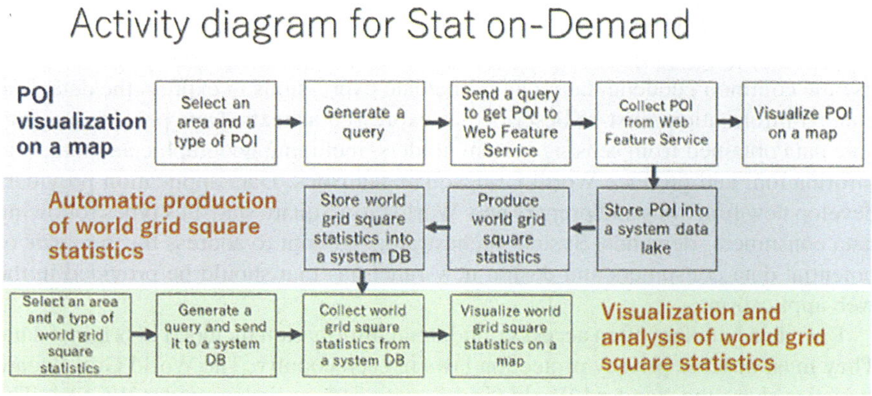

Fig. 7.1 Activity diagram for an on-demand system to produce Grid Square statistics

1. Store POI into a system data lake.
2. Produce World Grid Square statistics.
3. Store World Grid Square statistics into a system DB.

The visualization and analysis of World Grid Square statistics include four steps:

1. Select an area and a type of World Grid Square statistics.
2. Generate a query and send it to a system DB.
3. Collect World Grid Square statistics from a system DB.
4. Visualize World Grid Square statistics on a map.

Using the proposed procedure, we can automatically produce Grid Square statistics from POI and visualize them on a map. In the next section, we show a prototype visualization system implementing the proposed procedure as a data application of MESHSTATS [11].

7.4 Implementation

As an example of Web Feature Service, we focus on OpenStreetMap Overpass. OpenStreetMap (OSM) [12] provides point data. Users can extract point data regarding context information and geographical area using a query shown in Fig. 7.2. Data are provided through Web API and are formatted as JSON and XML [14].

We implemented the proposed system in MESHSTATS as a Web application [11]. Users can select a type of Grid Square statistics in a given area and a type of POI from a list of features. The system generates a query and sends it to collect POI from Web Feature Service. The system visualizes POI on a Grid Square map. Figure 7.3 shows a snapshot of the proposed Web application for visualizing Grid Square map, which shows selected World Grid Square statistics on a map. We can measure

```
[out:json][timeout:25];
// gather results
(
  // query part for: "temple"
  node["amenity"="toilets"](34.891667,135.575000,35.350000,135.875000);
  way["amenity"="toilets"](34.891667,135.575000,35.350000,135.875000);
  relation["amenity"="toilets"](34.891667,135.575000,35.350000,135.875000);
);
// print results
out body;
>;
out skel qt;
```

Search area

Keywords

An example of query to search "amenity"="toilets" near Kyoto City

```
http://overpass-api.de/api/interpreter
?data=%5Bout%3Ajson%5D%5Btimeout%3A25%5D%3B%
28node%5B%22amenity%22%3D%22toilets%22%5D%283
4%2E891667%2C135%2E575%2C35%2E35%2C135%2E87
5%29%3Bway%5B%22amenity%22%3D%22toilets%22%5
D%2834%2E891667%2C135%2E575%2C35%2E35%2C135
%2E875%29%3Brelation%5B%22amenity%22%3D%22toil
ets%22%5D%2834%2E891667%2C135%2E575%2C35%2E
35%2C135%2E875%29%3B%29%3Bout%3B%3E%3Bout%
20skel%20qt%3B%0A
```

Fig. 7.2 An example of Web API query for OpenStreetMap overpass

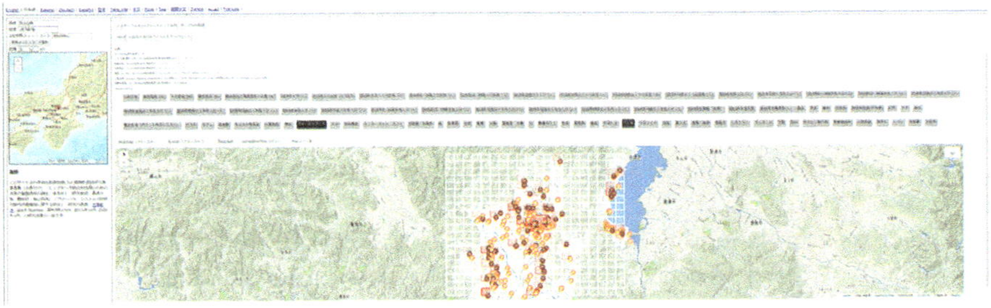

Fig. 7.3 Snapshots of the proposed system for visualizing Grid Square map and POI collected from Web Feature Service

regional characteristics from both point data of selected objects and their Grid Square statistics on a map.

We show an example of data analysis on Grid Square statistics produced by the proposed system. The correlation among Grid Square statistics is related to regional characteristics. This example uses Grid Square statistics for measuring similarity between cities [10].

Consider n types of properties such as $i = 1, \ldots, n$. The point data about the ith property contain $m(i)$ observations. The number of ith property at Grid Square k is denoted as $x(i)_k \quad (k = 1, \ldots, S)$, where S is the number of Grid Squares included in domain D contained in city s. We have

$$\sum_{k=1}^{S} x(i)_k = m(i) \quad (i = 1, \ldots, n).$$

We define a mean of the number of objects per Grid Square and its sample variance as

$$E[x(i)] = \frac{m(i)}{S},$$
$$V[x(i)] = \frac{1}{S}\sum_{k=1}^{S}\Big(x(i)_k - E[x(i)]\Big)^2.$$

We can compute cross-correlation coefficients for city s defined as

$$\rho_{i,j}(s) = \frac{Cov[x(i), x(j)]}{\sqrt{V[x(i)]}\sqrt{V[x(j)]}}, \tag{7.9}$$

where $Cov[x(i), x(j)]$ is variance-covariance matrix defined as

$$Cov[x(i), x(j)] = \frac{1}{S}\sum_{k=1}^{S}\Big(x(i)_k - E[x(i)]\Big)\Big(x(j)_k - E[x(j)]\Big).$$

We selected 17 objects ($n = 17$) as shown in Table 7.1 [13] and collected point data from OSM for ten cities shown in Table 7.2. We set a circle area at the specified position indicated by latitude and longitude with a 10 km radius to collect point data from OSM (Table 7.2).

The cross-correlation coefficients among 17 objects contain 17×17 elements. We quantify a difference between two cities by using a mean squared difference between two cross-correlation matrices at cities s and t:

$$K(s, t) = \frac{1}{n^2}\sum_{i=1}^{n}\sum_{j=1}^{n}\Big(\rho_{i,j}(s) - \rho_{i,j}(t)\Big)^2, \tag{7.10}$$

Table 7.3 shows mean squared differences computed from cross-correlation matrix for two cities.

Table 7.3 shows the mean squared difference among ten cities. We further apply hierarchical clustering analysis on the mean squared difference matrix. Figure 7.4

Table 7.1 Key-val expression for objects and their description

Key	Val	Objects
Shop	Clothes	A shop which primarily sells clothing
Shop	Bakery	A shop selling bread
Shop	Alcohol	A shop selling alcoholic drinks
Shop	Shoes	A shop selling shoes
Amenity	Restaurant	A restaurant sells full sit-down meals with servers
Amenity	Fast_food	A place concentrating on very fast counter-only service and take-away food
Shop	Convenience	A small local shop
Amenity	Pharmacy	A shop where a pharmacist sells medications
Shop	Confectionery	A shop selling sweets and candies
Shop	Massage	Massage shop
Amenity	Bank	A financial establishment
Amenity	Atm	A device that provides the clients of a financial institution with access to financial transactions
Shop	Hairdresser	A hairdressers or barbers shop
Tourism	Hotel	Hotel accomodations
Amenity	Toilets	A publicly accessible toilet
Amenity	School	A primary or secondary school (Pupils typically aged 6–18)
Amenity	Kindergarten	A place for looking after preschool children

Table 7.2 A list of cities and their location where point data ware collected from OSM

City name	Country	(Latitude, Longitude)
Yokohama	Japan	(35.428784, 139.646463)
Kyoto	Japan	(35.009129, 135.754807)
Osaka	Japan	(34.75003522, 135.4601448)
Tokyo	Japan	(35.68501691, 139.7514074)
Sapporo	Japan	(43.055248, 141.345505)
Roma	Italy	(41.89193, 12.51133)
Berlin	Germany	(52.520008, 13.404954)
Brisbane	Australia	(–27.46977, 153.025131)
Manila	Philippines	(14.599512, 120.984222)
Singapore	Singapore	(1.32225, 103.836401)

Table 7.3 Mean squared difference between two cross-correlation matrices for ten cities

City	Yokohama	Kyoto	Osaka	Tokyo	Sapporo	Roma	Berlin	Brisbane	Manila	Singapore
Yokohama	0.000000	2.619542	1.175946	1.971191	6.475579	3.144375	1.295019	3.316331	3.191194	1.711265
Kyoto	2.619542	0.000000	1.720845	1.683116	3.295287	2.349172	2.814103	3.014115	2.380269	3.260106
Osaka	1.175946	1.720845	0.000000	1.167970	5.333140	1.879794	1.657827	2.663148	2.395846	1.631623
Tokyo	1.971191	1.683116	1.167970	0.000000	5.188704	2.037375	1.908887	2.217978	1.992334	2.074744
Sapporo	6.475579	3.295287	5.333140	5.188704	0.000000	3.682643	6.617194	4.197021	2.983804	6.329284
Roma	3.144375	2.349172	1.879794	2.037375	3.682643	0.000000	2.707398	2.231983	1.869633	2.097515
Berlin	1.295019	2.814103	1.657827	1.908887	6.617194	2.707398	0.000000	3.538410	2.592064	1.321277
Brisbane	3.316331	3.014115	2.663148	2.217978	4.197021	2.231983	3.538410	0.000000	2.191660	2.876099
Manila	3.191194	2.380269	2.395846	1.992334	2.983804	1.869633	2.592064	2.191660	0.000000	2.387787
Singapore	1.711265	3.260106	1.631623	2.074744	6.329284	2.097515	1.321277	2.876099	2.387787	0.000000

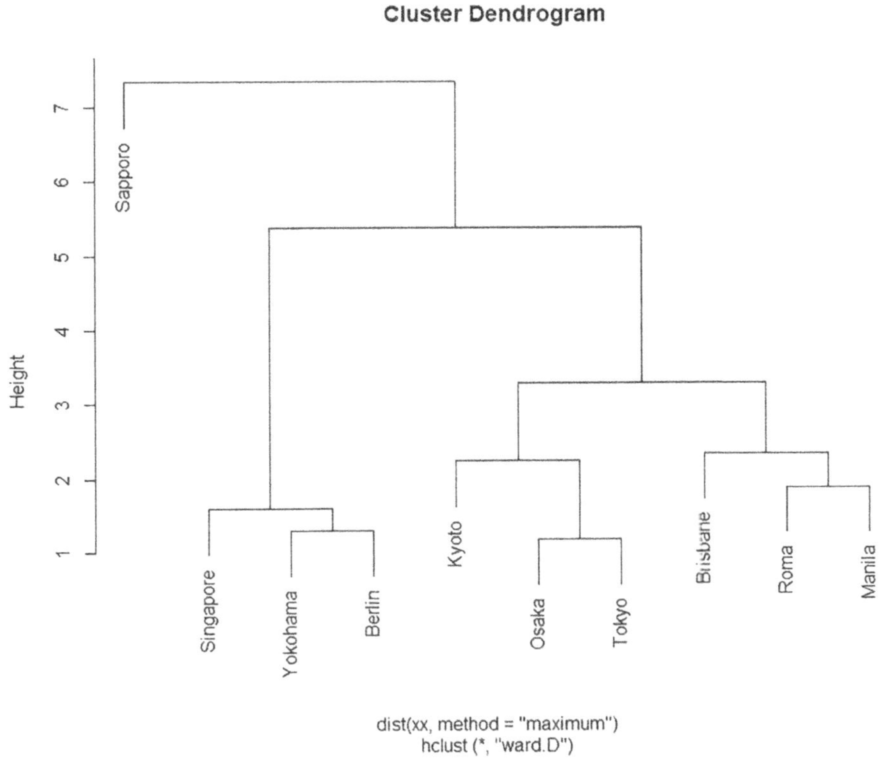

Fig. 7.4 Cluster dendrogram for the mean difference matrix for ten cities

shows the cluster dendrogram. From Fig. 7.4, we may interpret the similarity between cities as follows. Sapporo is the most different from other cities. Brisbane, Rome, Manila, Kyoto, Osaka, and Tokyo are in the same group. Singapore, Yokohama, and Berlin are in the other same group. We may characterize and classify cities from point data of various objects collected from OSM.

7.5 Extension

The descriptive accuracy of latitude/longitude information is relatively high. However, humans need assistance understanding and remembering latitude and longitude values quickly. Thus, geographical information such as a postal address and postal code is often used to express physical location in various real-world data types.

Postal addresses are expressed as a sequence of letters or characters, including numeric expressions with hierarchy. A postal code (postcode, PIN, or ZIP Code) is a

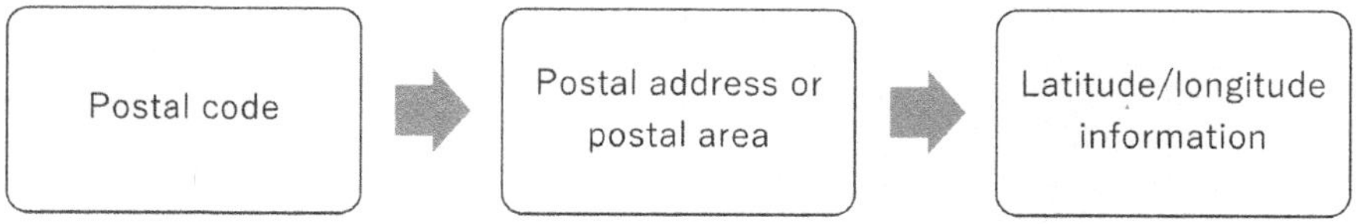

Fig. 7.5 Relations among postal codes, postal addresses or postal areas, and the latitude/longitude information

series of letters or digits. They are usually assigned to geographical areas. Moreover, some special codes are assigned to individual addresses or institutions that receive many emails, such as government agencies and commercial firms.

As shown in Fig. 7.5, postal addresses are converted into latitude and longitude information using a geocoding engine. The database can convert a postal code to a postal address. Therefore, we can convert postal codes and addresses into the latitude and longitude information and apply our proposed method explained in Sect. 7.3 for point data to data, including postal addresses and postal codes.

We propose data process flows for data including postal addresses and postal codes, modifying the data process flow for the latitude/longitude information shown in Fig. 7.1. Figure 7.6 shows the system activity, such as POI visualization on a map, automatic production of World Grid Square statistics, visualization, and analysis of World Grid Square statistics for postal addresses and postal codes.

We use data, including postal addresses and postal codes, to produce, visualize, and analyze Grid Square statistics. Users with tabular data, including postal codes or addresses, can convert their tabular data into point data (latitude and longitude information) based on postal codes and postal addresses in files uploaded by users to the Web service. The proposed data process flows shown in Fig. 7.6 can be realized. The Web system computes Grid Square codes from the latitude and longitude information in tabular data, aggregates statistics using Grid Square codes, and produces Grid Square statistics.

7.6 Conclusion

This chapter proposed a concept of Grid Square statistics on-demand. We designed the logic of the on-demand service based on GSBPM, which consists of specifying needs, designing, building, collecting, processing, analyzing, disseminating, and evaluating. We also considered architecture based on WGSDRA, including views by sensing system holder, data provider, data application provider, data consumers, and data orchestrator.

This chapter considered the requirements and designed an activity diagram for the on-demand service on Grid Square statistics, which can automatically produce Grid Square statistics from point data collected from OSM is based on user behavior.

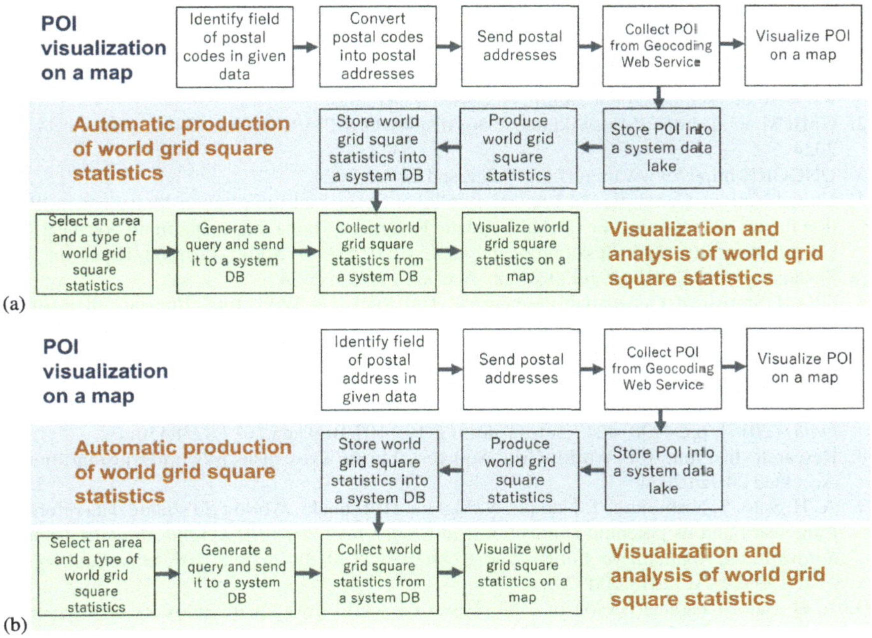

Fig. 7.6 Activity diagram for an on-demand system to produce Grid Square statistics for data including postal addresses (**a**) and postal codes (**b**)

We developed a data process algorithm consisting of data acquisition, collection, computation, production, and visualization. We implemented the system to produce Grid Square statistics from point-of-interests (point data) collected from the Open Street Map on demand. We prototyped a Web data application of Grid Square statistics in MESHSTATS. We visually showed the prototype of such an on-demand service with several visualization functions, such as Grid Square map visualization, cross-correlation, and spatial regression analysis. Moreover, we showed that our proposed data process flow can be extended to other geographical expressions such as postal addresses and postal codes.

We showed an example of data analysis to compare cities from a correlation matrix among objects computed from their Grid Square statistics. This data analysis is used to understand and evaluate the temporal differences between cities. The results may be used to consider strategy improvement in business and government organizations on Check and Action for the PDCA cycle in continuous improvement.

In future work, predicting users' interest from their data-searching behavior will be helpful. Moreover, users' similarity can be measured from their data-searching behavior. Such measurement can be used to categorize users who use Grid Square statistics through the Web Feature service about Grid Square statistics.

References

1. ISO/IEC 20546:2019, Information technology-Big data-Overview and vocabulary. https://www.iso.org/standard/68305.html. Accessed 9 Oct 2022
2. GSBPM v5.0. https://statswiki.unece.org/display/GSBPM/GSBPM+v5.0. Accessed 24 Jan 2024
3. UNGGRF. https://www.unggrf.org. Accessed 24 Jan 2024
4. United Nations Committee of Experts on Global Geospatial Information Management, Position Paper on Sustaining the Global Geodetic Reference Frame. https://ggim.un.org/meetings/GGIM-committee/11th-Session/documents/E-C.20-2021-7-Add-2_Position_Paper_on_Sustaining_the_GGRF_29Jul2021.pdf. Accessed 20 Jan 2024
5. Global Statistical Geospatial Framework, UN Statistics Wiki. https://unstats.un.org/wiki/display/GSGF/Global+Statistical+Geospatial+Framework. Accessed 25 Jan 2024
6. JISC, The Japanese National Grid Square codes, JIS X0410:2002. Accessed 24 Jan 2024
7. A.-H. Sato, S. Nishimura, H. Tsubaki, World grid square codes: Definition and an example of world grid square data, in *Conference: 2017 IEEE International Conference on Big Data (Big Data)* (2017), pp. 4238–4247. https://doi.org/10.1109/BigData.2017.8258450
8. Research Institute for World Grid Squares. https://www.fttsus.jp/worldgrids/en/library/. Accessed 24 Jan 2024
9. A.-H. Sato, S. Nishimura, T. Namiki, N. Makita, H. Tsubaki, World grid square data reference framework and its potential applications, in *Conference: 2018 IEEE 42nd Annual Computer Software and Applications Conference (COMPSAC)* (2018), pp. 398–409. https://doi.org/10.1109/COMPSAC.2018.00062.
10. A.-H. Sato, Characterization of cities based on world grid square statistics about specific properties, in *Conference: 2017 IEEE International Conference on Big Data (Big Data)* (2017), pp. 4228–4237. https://doi.org/10.1109/BigData.2017.8258449
11. MESHSTATS. http://www.meshstats.net/meshstats/. Accessed 24 Jan 2024
12. OpenStreetMap. https://openstreetmap.org/. Accessed 24 Jan 2024
13. OpenStreetMap Taginfo. https://taginfo.openstreetmap.org/. Accessed 24 Jan 2024
14. OpenStreetMap Overpass. http://overpass-turbo.eu/. Accessed 24 Jan 2024

Chapter 8
Case Studies

Abstract This chapter shows some examples of Grid Square statistics and their applications. Specifically, we review several studies on Grid Square statistics. Mainly using Japanese data and statistics, we consider some use cases for disseminating official statistics such as population census, measuring economic activities of tourism, natural disaster risk assessment, and an SDGs indicator (SDG 11.3.1). Moreover, we explain the concept of text Grid analysis and show examples of text Grid Square data analysis.

8.1 Use Cases and Scenarios

Big data analytics [1–9] requires advanced techniques, methods, tools, and applications to acquire, collect, process, compute, examine, and derive valuable insights from varied and large datasets (Big data) from different data sources. These data sets are often generated at very high speed with high variety. Grid Square statistics provide several valuable solutions to tackling challenges in Big data analytics.

Use cases of Grid Square statistics may be classified into the following six cases:

1. **Data linkage and data processing**: We can link different Grid Square statistics (linkage) and use them in operations, among other Grid Square statistics, to synthesize new Grid Square statistics from several original data types.
2. **Mapping**: We can visualize Grid Square statistics on a map to analyze our focal area.
3. **Data creation on given areas**: We can produce statistics on a given area by recalculating Grid Square statistics.
4. **Identifying effective areas**: Because Grid Squares make it easy to measure distances among themselves, Grid Square statistics can be used to calculate demand within a given distance.
5. **Defining observation areas**: Grid square statistics can define an area for collecting data or samples.
6. **Unit for numerical simulation**: Grid Square statistics can be used to conduct numerical simulations for a unit such as diffusion processes, percolation models, and migration processes.

A.-H. Sato and H. Tsubaki, *Evaluation Platform of Sustainability for Global Systems*,
https://doi.org/10.1007/978-981-97-2296-9_8

The list above shows data linkage and processing, mapping, data curation on given areas, identifying effective areas, and defining observation areas and units for numerical simulation as use cases for Grid Squares statistics. In the data linkage and data processing application, we can link different Grid Square statistics and use them in operations, among other Grid Square statistics, to synthesize new Grid Square statistics from several original data sources. In the mapping application, we can visualize Grid Square statistics on a map to analyze the focal area. In the data curation on given area applications, we can produce statistics on a given area by recalculating Grid Square statistics for that area. Grid Square statistics regarding distances and areas are used to identify effective area applications. In applications to define observation areas, Grid Square statistics can be used to determine an area for collecting data or samples. Finally, in the numerical simulation application, Grid Square statistics can conduct numerical simulations for a unit, such as diffusion processes, percolation models, and migration processes.

The following subsections summarize methods that can be used to analyze regional issues as a data-driven continuous improvement method based on Grid Square statistics.

8.1.1 Aggregation of Quantitative Grid Square Statistics in Arbitrary Regions

Comprehensive Grid Square statistics data products are provided, including Grid Square statistics for official statistics [10–12], commercial-based floating population [13–16], Wi-Fi log-based population density [17], and satellite data [18].

The Statistics Bureau of Japan provides Grid Square statistics about the Population Census, the Economic Census, and the Establishment and Enterprise Census [10] based on JIS X0410:2002.

Ministry of Land, Infrastructure, Transport, and Tourism also provides temporal development of the Japanese future population estimates as Grid Square statistics [12]. They produce Grid Square statistics about future population estimates based on cohort survival calculations from the Japanese Census population provided by the Statistics Bureau of Japan.

US Census provides Gridded Population Mapping data [19] in TIFF formats with 3 arc seconds × 3 arc seconds spatial resolution (approximately 100 m). US Census uses methods to estimate high-resolution gridded population based on a combination of census data and satellite image analysis. The gridded population is stored as a pixel of TIFF, and each pixel is placed by latitude and longitude on WGS84. Wang et al. produce a global gridded future population dataset at 30 arcseconds (approximately 1 km) spatial resolution with 5-year intervals for the period between 2010 and 2020 for 248 countries [20]. Schung et al. estimate German gridded population from earth observation data, building information collected from OpenStreetMap (OSM), and Census population [21].

Neki created Grid Square Data about the Bhutan Census Population and evaluated the advantages, disadvantages, and difficulties of the Grid Square format in Bhutan. He converts Bhutan's 2017 Population and Housing Census aggregated in Enumeration Area unit into the 6th level (approximately 125 m) World Grid Square statistics [22]. He concludes that using Grid Square data could improve the efficiency of spatial data analysis and data availability.

Grid Square statistics enable us to calculate various indicators for measuring sustainability by extracting and re-aggregating such Grid Square statistics in any area of the target region. For example, by re-aggregating the Census Grid Square statistics provided by the Statistics Bureau of Japan in small areas, age and gender information can be calculated for areas different from generally published aggregation units such as prefectures and cities. It is possible to calculate the population and number of single-person households. Furthermore, commercial human density Grid Square statistics make it possible to aggregate the resident population in any area at any time. Using Grid Square statistics from satellite data, measuring the green cover index of any region and temporal changes in land use rate is possible.

By aggregating the aggregation area in small areas such as railways, roads, and commercial facilities, it is possible to summarize the status of public transportation and land use using Grid Square code sets and calculate indicators.

8.1.2 Grid Square Statistics for Questionnaire Survey

The traditional method is to investigate local opinions through questionnaire surveys. A standard procedure is to aggregate survey respondents' opinions and use the statistics. Currently, questionnaire surveys are used to identify regional issues and solve them, but converting the data into grid square statistics makes it possible to capture the spatial bias of opinions in small regions.

Furthermore, combining with other grid square statistics makes it possible to analyze the correlation between location features and opinions. Understanding the spatial unevenness of regional issues allows us to know where the most problematic groups are located within regional issues and can also be used for gap analysis and policy analysis in continuous improvement.

Moreover, Grid Square statistics aggregated from questionnaire surveys can be used to design plans that consider the spatial arrangement of inputs when solving regional issues as part of the PDCA cycle and to design surveys when measuring effects.

When conducting a questionnaire survey by physical mail, it is easy to link the answers to the questionnaire survey with the postal code. Therefore, when conducting a general physical mail-based questionnaire survey in a region where solving regional issues is being considered, it is straightforward to produce grid square statistics for the questionnaire survey.

Furthermore, since online surveys are becoming more popular due to the recent high Internet penetration rate, a method that uses these surveys to collect postal codes as one of the questions is effective.

Since we should generally resolve regional issues within a limited scope, we can use Grid Square statistics to identify regional problems, policy analysis, and impact evaluation. We could produce Grid Square statistics by conducting physical mail or online surveys and can use them as evidence and knowledge for approaching regional issues.

8.1.3 Grid Square Statistics Produced from Image Analysis

By converting the image analysis index value of image information linked to location information into Grid Square statistics, the characteristics of places visually identified by people can be aggregated as Grid Square statistics and combined with other Grid Square statistics.

For example, we could produce Grid Square statistics by extracting and analyzing aeronautical satellite imageries and map images in a Grid Square, extracting the feature values of the images, and linking these with Grid Square codes.

It is possible to mechanically extract the spatial characteristics of a place in a scalable manner. Furthermore, extracting feature amounts from photo images or videos linked to location information and linking them with Grid Square codes makes it possible to produce Grid Square data products of image analysis information of photos and video images.

Nowadays, with the popularity of smartphones, if it is possible to collect photos and videos taken by people living in a local area by linking them to a Grid Square code on a server, it will automatically collect many images related to a location. Image analysis of video data linked to location, collected through such a system, may make it possible to manage and monitor data on regional issues confidentially.

Examples include apps taking and posting photos of utility poles and manholes, weather observation, and services that monitor the deterioration of public buildings such as embankments and parks.

In addition, it could be possible to build a Grid Square data infrastructure for real-time regional monitoring by linking videos captured by security cameras with location information and producing a Grid Square statistics data product based on the video image analysis results.

8.2 Data Analysis for Mobile Spatial Statistics

Grid Square statistics about population of flow [13–16] are produced from GPS point data collected from mobile devices via user application software or log data collected in mobile phone network operations [23, 24]. Grid Square statistics about population

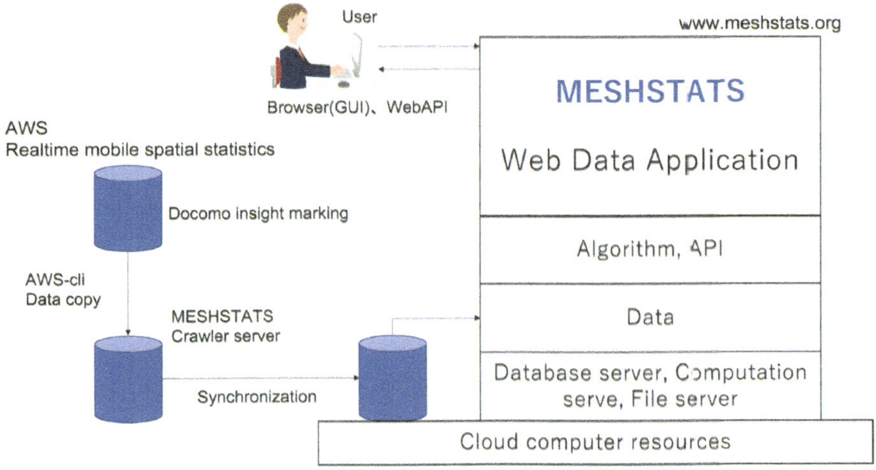

Fig. 8.1 A conceptual illustration of data application for real-time mobile spatial statistics

flow can be purchased from commercial data application providers and are used to monitor human activities in specific locations.

We construct Web API for real-time mobile spatial statistics [13] and conduct real-time population data analysis for specific sites during the Tokyo Olympics 2020 from 15 July 2021 to 13 August 2021. Figure 8.1 shows the structure of data application for real-time mobile spatial statistics. Since the Mobile Spatial Statistics data is a large amount of data, it is saved in the system as a compressed file, and when accessing the data, the data can be retrieved while partially expanding the necessary files.

The Web API can retrieve data by specifying time and location (latitude, longitude, and radius or a set of grid square codes). We analyze and visualize only the data required by extracting it through Web API.

We identified Grid Squares in advance from the location coordinates of the facility by using Grid Square codes. We converted location coordinates from the physical address or zip (postal) codes and computed Grid Square codes from the location coordinates. We observed the specified Grid Squares and drew the time series of population flow every hour. By measuring the normal 24 h activity status and the 24 h activity status on a specific date, we found the difference between the activity of a particular date and the regular activity.

Figure 8.2 shows a time series of the real-time population at the Tokyo Olympic main stadium from 15 July 2021 to 15 August 2021. We confirmed that the population increased by about 2,500 persons during the Tokyo Olympic 2020 opening ceremony between 20:00 and 24:00.

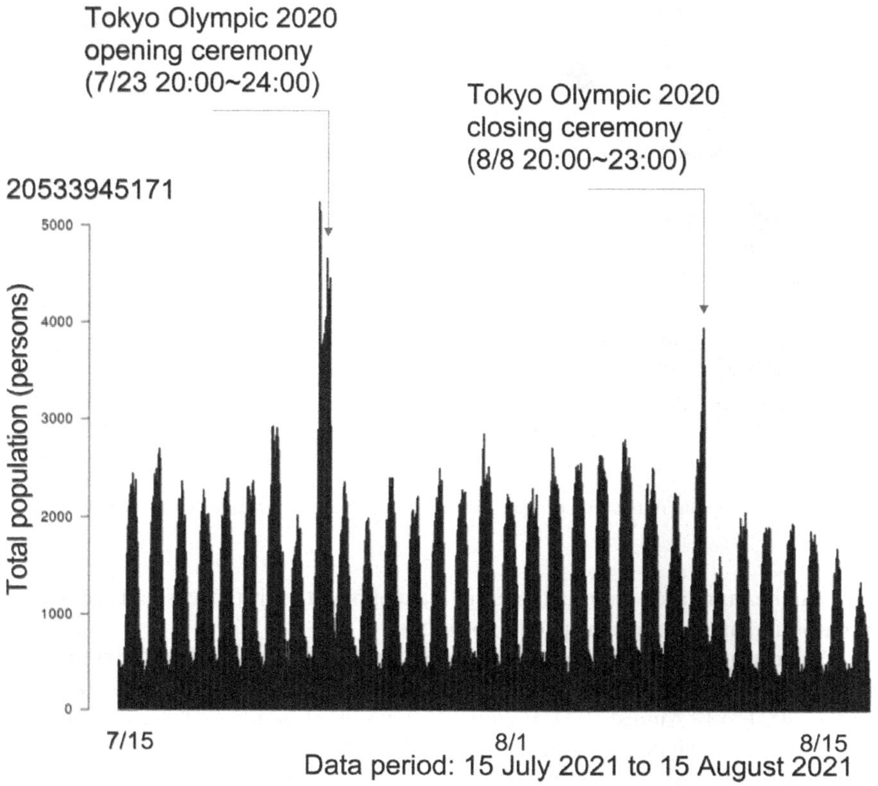

Fig. 8.2 A time series of the real-time population at Tokyo Olympic main stadium for a period from 15 July 2021 to 15 August 2021

8.3 Tourism

8.3.1 Tourism Statistics

Reference [25] produced such Grid Square data from micro-data of the Accommodation Survey in Japanese Tourism Statistics.

The Accommodation Survey in Japanese Tourism Statistics [26] is a quarterly survey conducted by the Japan Tourism Agency (Kankocho) of the Ministry of Land, Infrastructure, Transport and Tourism (MLIT). The survey collects data concerning accommodations and travel trends from several perspectives across the country and provides a fundamental database that captures the Japanese tourism sector and informs tourism policies. We can produce the third level Grid Square statistics data from micro-data of the Accommodation Survey based on Article 33 of the Statistics Act in Japan.

The Japanese accommodation statistics are based on whole sampling for hotels with more than ten labors, 1/3 random sampling for hotels with ten to five labors, and 1/9 random sampling for hotels with less than five labors.

There are four steps to produce Grid Square statistics from micro-data: (1) Data validation, (2) Geocoding, (3) Encoding, and (4) Totaling. The Japanese Tourism Statistics include

- the total and actual number of travelers,
- the total and actual number of foreign travelers per month,
- the actual number of used rooms (Question 7), and
- the number of foreign travelers per nationality.

We confirmed that foreign travelers stayed in various areas of Japan. Meanwhile, we confirmed that there are some highly dense areas. We can recognize the differences in travel patterns between people in China and the USA. We also confirmed that Chinese tourists visited western Japan and Hokkaido more than tourists from the USA. We further compare the number of tourists in each Grid Square.

Reference [25] investigates Japanese tourism patterns by using World Grid Square statistics produced from raw data of official tourism statistics. It also conducts regression analysis among World Grid Square statistics. We use the World Grid Square statistics about available hotel room opportunities and accommodations to estimate room occupancy rates.

8.3.2 Data Description

The data about hotel room availability are used to monitor hotel booking status all over Japan. We used a Web API (Application Programming Interface) to collect the data from a Japanese hotel booking site named Jalan. Jalan is one of the most popular hotel booking sites in Japan. They provide about 200 thousand available room opportunities offered by 1.8 thousand hotels.

Jalan Web service [27] provides interface for both hotel managers and consumers. The hotel managers offer their available rooms through Web-based interface. The consumers can book rooms from available rooms via the Jalan Web site. The third parties can even use Jalan Web API to build their own web services with hotel room opportunities provided through the Web API.

The data on available room opportunities in which two adults would be able to stay for one night are collected. The data on room opportunities collected through the Web API are stored as CSV files. The data collection date and the date of stay are the two types of dates. Usually, as the difference between the data collection date and the date of stay becomes longer, the amount of available room opportunities becomes larger.

Table 8.1 shows the contents included in the dataset. Each plan contains the data collection date, date of stay, regional sequential number, hotel identification number, hotel name, postal address, URL of the hotel web page, geographical position

Table 8.1 Data description of Jalan web API

No.	Description
1	Data collection date
2	Date of stay
3	Hotel identification number
4	Hotel name
5	Hotel name (Japanese kana)
6	Postal code
7	Address
8	URL
9	Latitude
10	Longitude
11	Room opportunity name
12	Meal availability
13	The latest best price
14	Price per night
15	The number of available room opportunities up to 10

(latitude and longitude), plan name, price, and the number of available room opportunities up to 10. Since the data contains regional information, we can analyze the hotel room prices by region.

8.3.3 Methods

We show a method to estimate the daily hotel occupancy ratio from available room opportunity data collected from the Internet booking site. Assume that we have the daily number of available plans $q(t)$ and the daily number of available rooms $r(t)$ offered by a hotel at date of stay t. The scatter plot between $q(t)$ and $r(r)$ can be fitted by monotonically increasing function in terms of $r(t)$.

Figure 8.3 shows the scatter plots between the daily number of rooms $r(t)$ and the daily number of plans $q(t)$.

Usually, the number of available rooms per room opportunity $r(t)/q(t)$ takes more than 1. However, if the number of available rooms is approaching 0, then the number of available rooms per room opportunity $r(t)/q(t)$ takes a value close to 1. In this case, the decreasing ratio of plans dq/dr is almost the same as the actual value of the decreasing room ratio. From this decreasing ratio, we can infer the actual number of rooms stored in the hotel from the available kinds of room opportunity $q*$ and the number of available rooms $r*$ satisfying $r * /p* = 1$ by $(dq/dr|_{q=q*})^{-1}q*$ since $q = dq/dr|_{q=q*}r$ is satisfied.

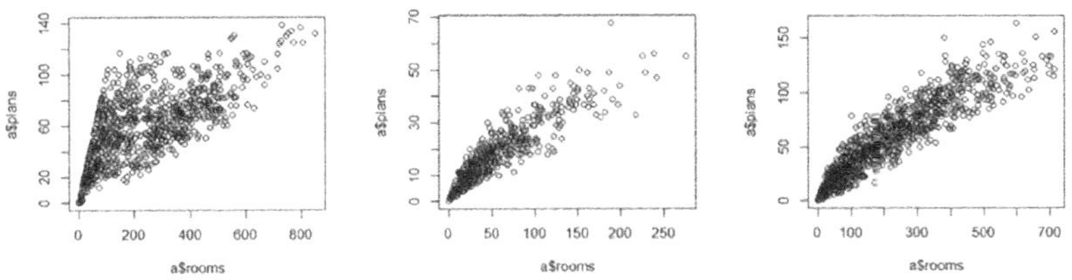

Fig. 8.3 Scatter plots between the daily number of room opportunities and the daily number of kinds of room opportunities for the third level Grid Squares

Assume that $q = cr^a$ is satisfied. This means that $q = c^r$ is satisfied at $q = q*$. This can be confirmed by a conditional regression analysis of $\ln q = \ln c + a \ln r$ for $r \leq r'$. If $a = 1$ is estimated then r' should be $r*$ and estimated c can be used as $dq/dr_{q=q*}$. Consequently, we can estimate the number of actually booked rooms from observable the number of available kinds of room opportunities $r(t)$ by using the relationship $r = c^{-1}q$. The daily occupancy ratio is inferred as

$$o(t) = 1 - \frac{c^{-1}q(t)}{\max_t\{q(t)\}}. \tag{8.1}$$

The monthly occupancy ratio can be computed from the daily occupancy ratio $o(t)$ such that

$$O(T) = \frac{1}{|m(T)|} \sum_{t \in m(T)} o(t), \tag{8.2}$$

where $m(T)$ is a set of dates included in month T and $|m(T)|$ represents the number of dates included in $m(T)$.

We define two dates for room reservation, such as the date of stay and data collection date, denoted as S and R, respectively. From this data, we compute the number of available room opportunities in Grid Square code w, denoted as $U_i(S, R)$. We assume that the daily number of available kinds of room opportunities in Grid Square code w is denoted as $P_i(S, R)$. $U_i(S, R)$ and $P_i(S, R)$ can be computed by accumulating the number of available room opportunities and geographical information (latitude and longitude) included in data collected from the Jalan Web API. We assume an area, a set of Grid Square codes Z. Thus, we can approximate the occupancy ratio in area Z as

$$\rho_Z(S, R) = 1 - \frac{\sum_{i \in Z} P_i(S, R)c^{-1}a^{-1}}{\sum_{i \in Z} \max_{S,R \in T} U_i(S, R)}, \tag{8.3}$$

where T shows the latest period including S and R. For example, T is set as the latest year. Moreover, if we have the number of potential rooms at which visitors can stay, which is called room capacity, C_i, then we can estimate the number of people who stay at hotels located in area Z as

$$N(S, R) = \sum_{i \in Z} C_i \rho_Z(S, R). \tag{8.4}$$

The room capacity is a slow variable based on government survey investigation as part of official statistics. In the case of Japan, we have official statistics from the accommodation survey conducted by Kankocho.

From the microdata of Japanese accommodation statistics, third level Grid Square statistics about room capacities are estimated by accumulating the number of room capacities in each accommodation in the same third level Grid Square. Assume that the total number of room capacities directly computed from microdata of accommodation survey in third level Grid Square i is denoted as γ_i. The number of hotels in third level Grid Square i is denoted as N_i and the number of answers in third level Grid Square i is denoted as n_i. Then the room capacity in third level Grid Square i is estimated as

$$C_i = \gamma_i N_i / n_i. \tag{8.5}$$

Using the procedure, we can approximate the daily hotel room occupancy rate and the number of visitors in the given area Z.

8.3.4 Empirical Analysis

The daily number of room opportunities in each 3rd-level Grid Square is computed from data collected from the Jalan Web API. Figure 8.4 shows the daily hotel room occupancy rate in Kyoto from 22 July 2017 to 5 December 2017. We found that there is weekly seasonality in Fig. 8.4. Typically, the daily hotel room occupancy rates

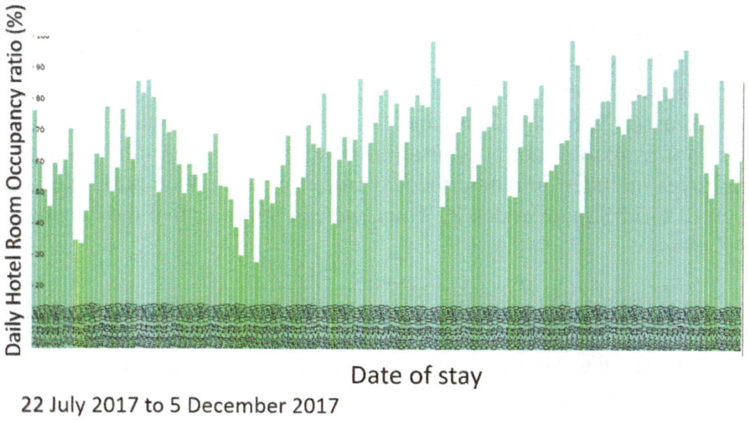

Fig. 8.4 The daily hotel room occupancy rates in Kyoto, Japan, from 22 July 2017 to 5 December 2017

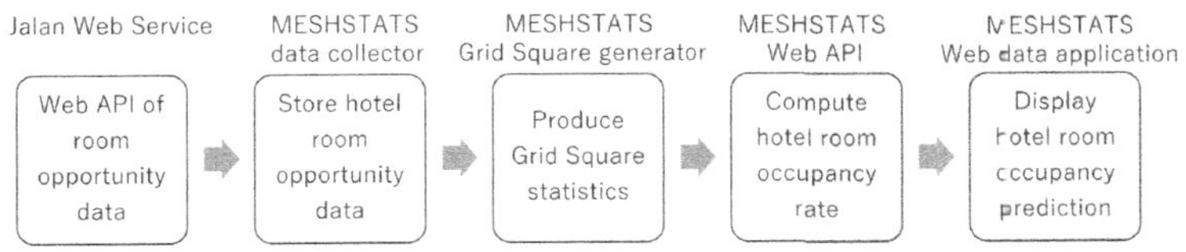

Fig. 8.5 Conceptual illustration to estimate hotel coom occupancy rates from room opportunity data collected from a hotel booking site

take the local maxima on Saturday and the local minima on Sunday. From Sunday to Saturday, the daily hotel room occupancy rate eventually increases. Moreover, the summer holiday and autumn seasons were highly demanded in Kyoto.

The function to compute daily hotel occupancy rate has been implemented in a Cloud-based data analytics platform, MESHSTATS. Figure 8.5 displays a conceptual illustration of procedures to approximate daily hotel occupancy rates. The procedure is as follows.

1. The primary hotel room opportunity data is automatically collected through Jalan Web API.
2. The primary data about hotel room opportunities are stored.
3. Grid Square statistics about the number of available room opportunities are produced.
4. Daily hotel room occupancy rates are computed via MESHSTATS Web API.
5. By differentiating areas, daily hotel room occupancy rates in different regions are computed and displayed through a data web application.

We can estimate daily room occupancy rates in a Grid Square selected as a set of World Grid Square codes from all over Japan.

8.3.5 Discussion

This relationship between the daily number of room opportunities and the daily number of room opportunity types shows some hotel and area characteristics. If a hotel introduces various room opportunities, then the increasing ratio of the number of room opportunity types in terms of the number of available rooms becomes high. Typically, hotels introduce many room plans to diversify their services. Such strategies strongly depend on regions.

Our method is one of the approximation techniques used to compute the daily hotel room occupancy ratio from available public information about an Internet booking site. The critical point of our method is how to infer a lower bound of the daily hotel room occupancy ratio from available public information. The highest hotel room occupancy ratio corresponds to no available rooms. However, the lowest hotel room occupancy ratio is unclear since the lower bound corresponds to the maximum available room opportunities. However, usually, this value is still being determined.

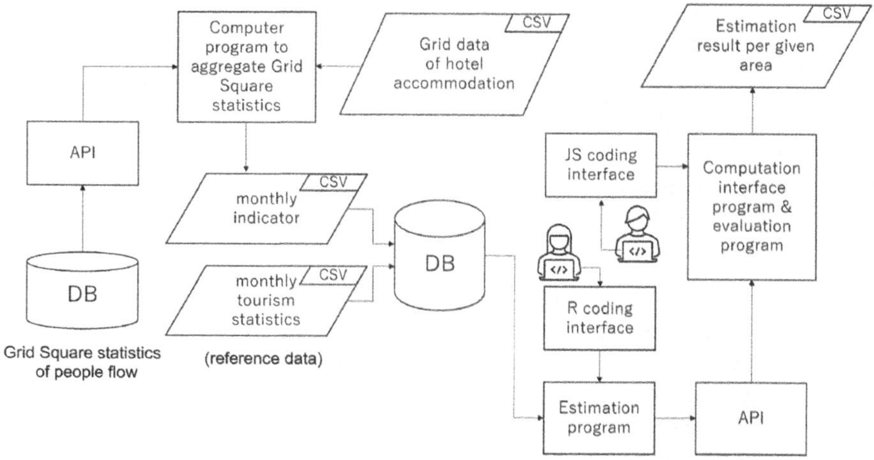

Fig. 8.6 A schematic diagram to estimate the total number of overnight stays at hotel accomodation per given area (prefecture, city and town)

An approximation method was proposed to infer the lowest value of the daily hotel room occupancy ratio.

Moreover, we can estimate the population who live in places other than where they stay during midnight from Grid Square statistics of floating population shown in Sect. 8.2. Figure 8.6 shows a schematic diagram to estimate the total number of overnight stays at accommodation by aggregating Grid Square statistics of the number of people who live in places other than where they stay during midnight in Grid Squares. The inputs are Grid Square statistics of floating population, Grid data of hotel accommodation, and monthly tourism statistics as reference data. The output estimates the total number of overnight stays at hotel accommodations per given area (prefecture, city, and town) with evaluation indicators.

8.4 Risk Assessment of a Natural Disaster

Using the Japanese National Grid Square statistics, successive studies have measured economic activities [28, 29] and assessed risks of natural disasters [30–32].

We mapped Grid Square statistics, enabling users to confirm regional dependence of World Grid Square statistics to measure socioeconomic activities and natural disaster hazards.

In general, risk is assessed as multiplication among hazard, vulnerability, and exposed value. In the case of risk of loss of lives at a Grid Square c, risk $R(c)$ represents an expected value of loss of lives, defined as

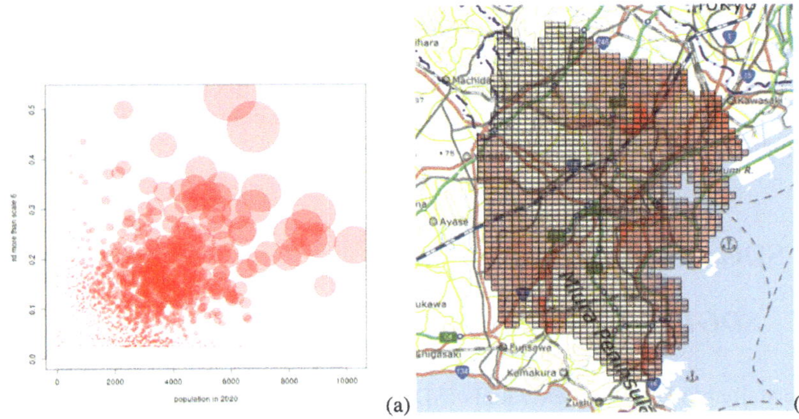

Fig. 8.7 **a** Scatter plots between the 1/2-level Grid Square statistics of seismic hazards more than scale six within 30 years in Yokohama and 1/2-level Grid Square statistics of population census in 2020. The size of the circle represents the physical exposure value. **b** 1/2-level Grid Square map of physical exposure between seismic hazards more than scale six and population census in 2020. Color depth represents physical exposure value

$$R(c) = F(c) \times Vul(c) \times Pop(c), \tag{8.6}$$

where $F(c)$ [events/year] (hazard) represents the frequency of a natural disaster, $Vul(c)$ is vulnerability, having a value ranging from 0 to 1, and $Pop(c)$ [persons/year] expresses the population exposed to the natural disaster. Specifically, the worst-case scenario ($Vul = 1$) is called physical exposure, which is defined as

$$PhExp(c) = F(c) \times Pop(c). \tag{8.7}$$

In Japan, we have Grid Square statistics of seismic hazards published by the National Research Institute for Earth Science and Disaster Resilience (NIED) [33], which is called Japan Seismic Hazard Information Station (J-SHIS) [30]. We computed 1/2-level Grid Square statistics of physical exposure between census population in 2020 published by Statistics Bureau of Japan [10] and 30-year probability for more than scale six earthquake to hit Yokohama City, Kanagawa Prefecture.

We identified 1,797 Grid Squares in Yokohama City and extracted seismic hazards more than scale six and population for identified 1/2-level Grid Squares. The total population in Yokohama City is 3,884,159, and total physical exposure is estimated at 15,463 persons/year.

Figure 8.7a shows scatter plots between the 1/2-level Grid Square statistics of seismic hazards more than scale six published by J-SHIS and 1/2-level Grid Square statistics of population census in 2020. We found that there are Grid Squares with high hazards and high physical exposure values. Figure 8.7b shows a Grid Square map of physical exposure between more than scale six seismic hazards and the population census in 2020.

It is found that some places in Yokohama City have high physical exposure values. We confirmed that several 1/2-level Grid Squares contain a significant risk of earthquakes against the population.

8.5 SDGs Indicator

8.5.1 SDG11.3.1

The Sustainable Development Goals (SDGs) [34] have been set as the core of the 2030 Agenda for Sustainable Development adopted by the United Nations Summit in September 2015. The SDGs consist of 17 goals and 169 targets, with 248 indicators (231 excluding duplicates) to measure activities related to SDGs with high-quality, accessible, timely, and disaggregated data.

Under Goal 11 "Sustainable cities and communities", aiming at participatory, integrated and sustainable human settlement planning and management for Target 11.3, there is Indicator 11.3.1 "the ratio of Land Consumption Rate to Population Growth Rate (LCRPGR)." SDG11.3.1 of the Custodian agency is the United Nations Human Settlements Programme (UN-Habitat), and according to the metadata published by the UN Habitat [35], SDG11.3.1 is measured as the ratio of Land Consumption Rate to Population Growth Rate (LCRPGR), which is defined as the ratio of Land Consumption Rate (LCR) to Population Growth Rate (PGR).

PGR_t is the change of population in a defined region (country, city, etc.) per year over a given period, and expressed as the natural logarithm of the ratio of population at the end of the period to the population at the beginning of the period divided by the number of years in the period. LCR_t is the change of land occupied by a city/urban area per year over a given period, and expressed as the ratio of the change of the land occupied by a city/urban over the period to the land occupied by a city/urban at the beginning of the period divided by the number of years in the period.

$$PGR_t = \frac{\ln(Pop_{t+n}) - \ln(Pop_t)}{y}, \tag{8.8}$$

$$LCR_t = \frac{\ln(Urb_{t+n}) - \ln(Urb_t)}{y}, \tag{8.9}$$

$$LCRPGR_t = \frac{LCR_t}{PGR_t}, \tag{8.10}$$

where Pop_t is the total population of the target region in the past year and Pop_{t+n} is the total population of the target region in the present year. Urb_t is the total area of city/urban area in the past year, Urb_{t+n} is the total area of city/urban area in the present year. y is the duration between the past and present years. $LCRPGR_t$ requires population data and urban and city area data. The metadata recommends the use of gridded population datasets and land cover datasets estimated from medium-to-high spatial resolution satellite data.

8.5.2 Grid Square Statistics of SDG11.3.1

When the third level Grid Square statistics of city ratio in year t with parccel indicated by the Grid Square w_i is $p_{t,i}$ and the area classified as city in year t from the defition of Eq. (8.10) is $m_t(w_i)$, $LCR_t(w_i)$ in Grid Square w_i is calculated from

$$LCR_t(w_i) = \frac{\ln m_{t+n}(w_i) - \ln m_t(w_i)}{y}. \tag{8.11}$$

Assuming that the area of Grid Square w_i is $A(w_i)$, one has

$$m_t(w_i) = A(w_i)p_{t,i}. \tag{8.12}$$

Substituting Eq. (8.12) into Eq. (8.11) can be rewritten as

$$LCR_t(w_i) = \frac{\ln A(w_i)p_{t+n,i} - \ln A(w_i)p_{t,i}}{y} = \frac{\ln p_{t+n,i} - \ln p_{t,i}}{y}. \tag{8.13}$$

Equation (8.13) implies that is is possible to calculate the LCR for Grid Square w_i using logarithm of the city ratio computed from gridded land cover data.

As shown in Sect. 4.3 (see Chap. 4), we may evaluate confidential interval for Gird Square statistics. The 95% confidence interval of the population LCR is calculated as

$$\tilde{LCR}_t(w_i) - \frac{1}{y}\Big(\frac{\delta\alpha_{t+n,i}}{p_{t+n,i}} + \frac{\delta\alpha_{t,i}}{p_{t,i}}\Big) \le LCR_t(w_i) \le \tilde{LCR}_t(w_i) + \frac{1}{y}\Big(\frac{\delta\alpha_{t+n,i}}{p_{t+n,i}} - \frac{\delta\alpha_{t,i}}{p_{t,i}}\Big), \tag{8.14}$$

where $\delta\alpha_{t+n,i} = 1.96\sqrt{\frac{p_{t+n,i}(1-p_{t+n,i})}{N_{t+n,i}}}$ and $\delta\alpha_{t,i} = 1.96\sqrt{\frac{p_{t,i}(1-p_{t,i})}{N_{t,i}}}$. Then,

$$\tilde{LCR}_t(w_i) - \frac{1.96}{y}\left(\sqrt{\frac{1-\tilde{p_{t+n,i}}}{n_{t+n,i}}} + \sqrt{\frac{1-\tilde{p_{t,i}}}{n_{t,i}}}\right) \le LCR_t(w_i) \le \tilde{LCR}_t(w_i)$$
$$+ \frac{1.96}{y}\left(\sqrt{\frac{1-\tilde{p_{t+n,i}}}{n_{t+n,i}}} + \sqrt{\frac{1-\tilde{p_{t,i}}}{n_{t,i}}}\right), \tag{8.15}$$

Table 8.2 Confusion matrix for city/non-city binary classification

Confusion matrix		Validated b	
	a/b	City ($x = 1$)	Non-city ($x = 0$)
Classified a	City ($x = 1$)	TP_t	FP_t
	Non-city ($x = 0$)	FN_t	TN_t

where the sample LCR is expressed by

$$L\tilde{C}R_t(w_i) = \frac{\ln(n_{t+n,i}/N_{t+n,i}) - \ln(n_{t,i}/N_{t,i})}{y}, \tag{8.16}$$

where $N_{t,i}$ and $n_{t,i}$ are the total pixels in Grid Square w_i identified by the land cover gridded data and the number of pixels identified as city or urban in Grid Square w_i, respectively.

We apply a correction method by confusion matrix of classification error. Correct classification error of city coverage ratios in the city/non-city binary classification problem by calculating the conditional probability $R(a|b)$ from a 2×2 confusion matrix (see Table 8.2) for city/non-city binary classification. Let us calculate 95% confidence interval of LCR which is corrected by the city/non-city classification error for each third level Grid Square.

Let $p_{t,i}$ be the true city population proportion of Grid Square w_i in year t, and define $R_t(a|b)$ as the conditional probability that actual land cover (validated) is $b = \{1(\text{city}), 0(\text{non-city})\}$, but land cover classification data (classified) is $a = \{1(\text{city}), 0(\text{non-city})\}$. From the confitional probability definition,

$$R_t(a|b) = \frac{p_{AB,t}(a,b)}{p_{B,t}(b)}. \tag{8.17}$$

The point estimate by sample proportions of the conditional probability $R_t(a|b)$ is calculated as

$$R_t(0|0) = \frac{TN_t}{FP_t + TN_t},$$
$$R_t(1|0) = \frac{FP_t}{FP_t + TN_t},$$
$$R_t(0|1) = \frac{FN_t}{TP_t + FN_t},$$
$$R_t(1|1) = \frac{TP_t}{TP_t + FN_t}.$$

From Eq. (8.17),

$$p_{A,t}(a) = \sum_{b=\{0,1\}} p_{AB,t}(a,b) = \sum_{b=\{0,1\}} R_t(a|b)p_{B,t}(b) \tag{8.18}$$

is formalized. Therefore, especially, one has

$$p_{A,t}(1) = \sum_{b=\{0,1\}} R_t(a|b)p_{B,t}(b) = R_t(1|1)p_{B,t}(1) + R_t(1|0)p_{B,t}(0). \tag{8.19}$$

Then $p_{B,t}(1) = \hat{p}_{t,i}$ and $p_{B,t}(0) = 1 - \hat{p}_{t,i}$. Thus, the sample city proportion $\tilde{p}_{t,i} = n_{t,i}/N_{t,i} = p_A, t(1)$ calculated in the situation. Since the calculation error is included in the observation, one has

$$\tilde{p}_{t,i} = R_t(1|1)\hat{p}_{t,i} + R_t(1|0)(1 - \hat{p}_{t,i}). \tag{8.20}$$

Therefore, one obtains

$$\hat{p}_{t,i} = \frac{\tilde{p}_{t,i} - R_t(1|0)}{R_t(1|1) - R_t(1|0)}. \tag{8.21}$$

Here, only the Grid Square that is $\tilde{p}_{t,i} \geq R_t(1|0)$ is subject to estimation. Setting $a_{t,i} = \tilde{p}_{t,i}$, $b_t = R_t(1|0)$, and $c_t = R_t(1|1)$, we have the The sample LCR is calculated as

$$L\tilde{C}R_t(w_i) = \frac{\ln\left(\frac{a_{t+n,i}-b_{t+n}}{c_{t+n}-b_{t+n}}\right) - \ln\left(\frac{a_{t,i}-b_t}{c_t-b_t}\right)}{y}. \tag{8.22}$$

The 95% confidence interval for population LCR is

$$\begin{aligned} L\tilde{C}R_t(w_i) - \frac{1}{y}\left(\frac{\delta a_{t+n,i} + \delta b_{t+n}}{a_{t+n,i} - b_{t+n}} + \frac{\delta c_{t+n} + \delta b_{t+n}}{c_{t+n} - b_{t+n}} + \frac{\delta a_{t,i} + \delta b_t}{a_{t,i} - b_t} + \frac{\delta c_t + \delta b_t}{c_t - b_t}\right) \leq LCR_t(w_i) \\ \leq L\tilde{C}R_t(w_i) + \frac{1}{y}\left(\frac{\delta a_{t+n,i} + \delta b_{t+n}}{a_{t+n,i} - b_{t+n}} + \frac{\delta c_{t+n} + \delta b_{t+n}}{c_{t+n} - b_{t+n}} + \frac{\delta a_{t,i} + \delta b_t}{a_{t,i} - b_t} + \frac{\delta c_t + \delta b_t}{c_t - b_t}\right), \end{aligned} \tag{8.23}$$

where $\delta a_{t,i}$, δb_t, and δc_t are 95% confidence intervals of $\tilde{p}_{t,i}$, $R_t(1|0)$, and $R_t(1|1)$, respectively

We computed LCR (third level Grid Square statistics) from the JAXA HRLULC [37] and PGR (third level Grid Square statistics) from third level Grid Square statistics of MIC Population Census [10]. The calculation algorithm is implemented in Web data application on MESHSTATS. Figure 8.8 shows a Web data application [36] for calculating and visualizing SDG11.3.1(LCRPGR) indicators. The map shows third level Grid Square statistics about LCRPGR indicators and the table expresses indicators aggregated by administrative division.

SDG11.3.1 (LCRPGR)

JP-14 Kanagawa Prefecture

LCR

LCR | LCR2 | PGR | Download as a csv file

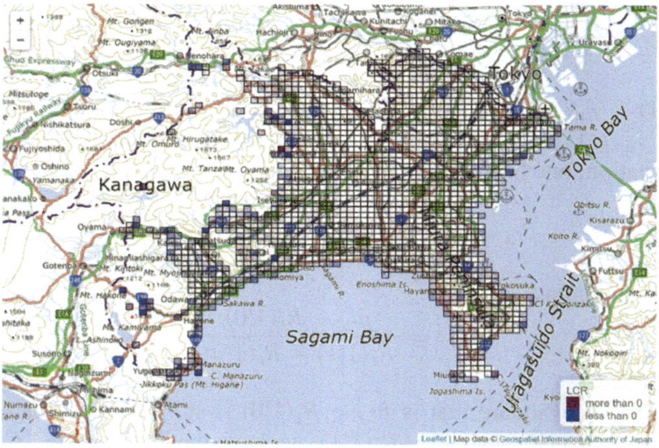

Sapporo City | Sendai City | Saitama City | Chiba City | Kawasaki City | Yokohama City | Sagamihara City | Niigata City | Shizuoka City | Hamamatsu City | Nagoya City | Kyoto City | Osaka City | Sakai City | Kobe City | Okayama City | Hiroshima City | Kitakyushu City | Fukuoka City | Kumamoto City | JP-01 Hokkaido Prefecture | JP-02 Aomori Prefecture | JP-03 Iwate Prefecture | JP-04 Miyagi Prefecture | JP-05 Akita Prefecture | JP-06 Yamagata Prefecture | JP-07 Fukushima Prefecture | JP-08 Ibaraki Prefecture | JP-09 Tochigi Prefecture | JP-10 Gunma Prefecture | JP-11 Saitama Prefecture | JP-12 Chiba Prefecture | JP-13 Tokyo Prefecture | JP-14 Kanagawa Prefecture | JP-15 Niigata Prefecture | JP-16 Toyama Prefecture | JP-17 Ishikawa Prefecture | JP-18 Fukui Prefecture | JP-19 Yamanashi Prefecture | JP-20 Nagano Prefecture | JP-21 Gifu Prefecture | JP-22 Shizuoka Prefecture | JP-23 Aichi Prefecture | JP-24 Mie Prefecture | JP-25 Shiga Prefecture | JP-26 Kyoto Prefecture | JP-27 Osaka Prefecture | JP-28 Hyogo Prefecture | JP-29 Nara Prefecture | JP-30 Wakayama Prefecture | JP-31 Tottori Prefecture | JP-32 Shimane Prefecture | JP-33 Okayama Prefecture | JP-34 Hiroshima Prefecture | JP-35 Yamaguchi Prefecture | JP-36 Tokushima Prefecture | JP-37 Kagawa Prefecture | JP-38 Ehime Prefecture | JP-39 Kochi Prefecture | JP-40 Fukuoka Prefecture | JP-41 Saga Prefecture | JP-42 Nagasaki Prefecture | JP-43 Kumamoto Prefecture | JP-44 Oita Prefecture | JP-45 Miyazaki Prefecture | JP-46 Kagoshima Prefecture | JP-47 Okinawa Prefecture

Type	adjusted LCR	minimum LCR	maximum LCR	95% CI	PGR	adjusted LCRPGR	minimum LCRPGR	maximum LCRPGR
old definition of LCR	0.0184	0.0078	0.0289	0.0211	0.0018	10.0242	4.2586	15.7898
new definition of LCR	0.0192	0.0077	0.0308	0.0232	0.0018	10.4989	4.1788	16.8191

pixel_count_before	37860000
built.up_count_before	11717771
pixel_count_after	4207000
built.up_count_after	1413718
population_before	9356854
population_after	9442973

Data sources

1. JAXA High-Resolution Land Use and Land Cover Map Product(2006-2011), (2014-2016)
2. 3rd level grid square statistics of census population (2000), Statistics Bureau of Japan, 3rd level grid square statistics of census popuation (2010), Statistics Bureau of Japan

Fig. 8.8 Grid Square map on SDG11.3.1 (LCRPGR) indicator shown in the web data application [36] and indicators aggregated by Kanagawa Prefecture

8.6 Text Grid Square Analysis

8.6.1 Concept

Using human-understandable text information as the data source for Grid Square statistics, we can produce Grid Square statistics for word frequencies and perform spatial analysis on Grid Square statistics for each word.

We need to distinguish the characteristics of places within a limited range in small regions for regional issues. It is convenient to analyze Grid Square statistics based on human-recognizable words to identify regional characteristics. Text Grid Square analysis become an effective method when we have text data with geographical information.

Scenarios in which we introduce the text Grid Square analysis to solve regional issues include producing text Grid Square statistics based on the text of written responses in the questionnaire above survey and analyzing text Grid Square statistics from both spatial and temporal perspectives. Text Grid Square analysis can effectively identify and visualize regional characteristics from words at a given Grid level. It can be an effective method to analyze written responses that are difficult to utilize in questionnaire surveys.

Furthermore, we can specify text embedded in real space when we conduct text Grid Square analysis using information sent by organizations, facilities, and companies from their web pages and Grid Square codes identified from their physical location information. It becomes possible to capture its semantic information as Grid Square statistics and use it to measure space.

8.6.2 Data Sources

We produced Grid Square statistics from text with location information as a data source. We collected 7,997,974 tweets with latitude and longitude from 17 December 2020 to 10 April 2023 receiving permission from Twitter Inc.

We can conduct spatial text analysis to characterize a place by natural language. In particular, identifying the Grid Square code (location) from the text is highly useful. To do so, we extract the words contained in the sentences with the Grid Square code from text data corresponding to location information.

We show some examples with a sentence with (a) an address, (b) a postal (or zip) code, and (c) latitude and longitude in Table 8.3.

Geographical information (latitude and longitude) can be calculated from an address or zip code by geocoding engine, and latitude and longitude can be converted into a Grid Square code. Furthermore, we can decompose test sentences into words by using a morphological analysis engine.

Table 8.3 (a) Examples of a set of location and text for (a) address, (b) a postal (zip) code, and (c) geographical position (latitude and longitude)

(a)	Address	Text
	22-2 Seto, Kanazawa-ku, Yokohama City, Kanagawa Prefecture, Japan	I found an interesting book
(b)	Postal code	Text
	236-0027	I found an interesting book
(c)	Geographical information [latitude, longitude]	Text
	[35.33446, 139.6171122]	I found an interesting book

Table 8.4 Sets of Grid Square codes and words

2052397499334	I found an interesting book
2053397499335	I ate a dish of steak
2052397499331	The school provides interesting classrooms
2052397500112	I know a list of good restaurants

As a result, by combining geocoding and morphological analysis, we can generate sets of Grid Square codes as discretized positions and words as morphological analyzed text data (see Table 8.4).

8.6.3 Bipartite Graph Representation and Cooccurence Matrix

As shown in Fig. 8.9, sets of Grid Square codes and words imply that a bipartite graph representation can be constructed between the Grid Square codes and the words with their frequency of occurrence as a weight.

In this case, the vertical vector $\mathbf{b}_{m1} = [1514 \cdots 4]^T$ means the frequency distribution in terms of the word for World Grid Square code 2052397500112. The horizontal vector $\mathbf{a}_{w4} = [4 \cdots 82]$ means the frequency distribution in terms of the Grid Square for word "stake."

Suppose that $W(m, w)$ is denoted as the cooccurrence matrix (two-dimensional contingency table) between the Grid Square code $m \in Mesh$ and the word $w \in Word$. Here, let $Mesh$ and $Word$ be a set of Grid Square codes and words contained in the text sentence with location information, respectively.

Then, a joint probability $p(m, w)$ can be approximately calculated as

$$p(m, w) = \frac{W(m, w)}{\sum_{m' \in Mesh} \sum_{w' \in Word} W(m', w')} \quad (8.24)$$

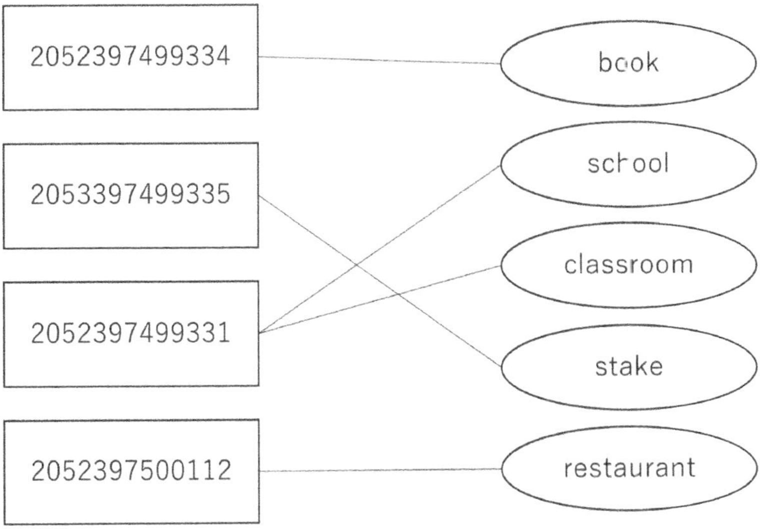

Fig. 8.9 Bipartite graph representation between Grid Squares and words

In addition, the marginal probabilities for Grid Square code m and word w are given by

$$q(m) = \sum_{w \in Word} p(m, w), \tag{8.25}$$

$$r(w) = \sum_{m \in Mesh} p(m, w). \tag{8.26}$$

The similarity between a Grid Square code and words can be defined by using marginal probabilities $q(m)$ and $r(w)$, respectively.

8.6.4 Similarity Measure

In fact, various similarity scales are available, but here we focus on Jensen-Shannon divergence [38] as an example. Jensen Shannon divergence is a subjective weight ($\pi_1 + \pi_2 = 1$) for two probabilities r_1 and r_2.

$$JS(r_1, r_2) = H[\pi_1 r_1 + \pi_2 r_2] - \pi_1 H[r_1] - \pi_2 H[r_2], \tag{8.27}$$

where $H[p]$ is Shannon entropy defined as

$$H[p] = -\sum_{x \in A} p(x) \ln p(x), \tag{8.28}$$

where A be the event of interest and we define $0 \ln 0 = 0$. The Jensen-Shannon divergence is also given by using Kallback-Leibler divergence (relative entropy) defined as

$$D_{KL}(p||q) = \sum_{xinA} p(x) \ln \frac{p(x)}{q(x)}. \tag{8.29}$$

Then, the Jensen-Shannon divergence is defined as

$$JS(r_1, r_2) = \frac{1}{2} D_{KL}\left(r_1 || \frac{r_1 + r_2}{2}\right) + \frac{1}{2} D_{KL}\left(r_2 || \frac{r_1 + r_2}{2}\right). \tag{8.30}$$

From the features of the Jensen-Shannon divergence,

- Two probabilities r_1 and r_2 coincides if and only if $JS(r_1, r_2) = 0$,
- $JS(r_1, r_2)$ takes the inequality $0 \leq JS(r_1, r_2) \leq \ln 2$,

we may defined an indicator of the correlation between two probabilities by standardized Jensen-Shannon divergence:

$$\rho = 1 - \frac{JS(r_1, r_2)}{\ln 2}, \tag{8.31}$$

where one clearly has $0 \leq \rho \leq 1$. Moreover, the two probabilities do not coincide $\rho = 0$, and the two probabilities coincide when $\rho = 1$. We can use ρ to compare two probabilities.

8.6.5 Examples

As an example of data analysis, we used tweets with location information for the period from 17 December 2020 to 19 April 2023, with 6,705,374 records, 364,886 words, 18,102 third level Grid Squares, 39,677 fourth-level Grid Squares, and 78,069 fifth-level Grid Squares.

The pipeline from data collection to data visualization is as follows:

- Collect tweets with location information in JSON format
- Identify the Grid Square code from the collected tweets with location information and save the morphological analysis result data as a CSV file.
- Search based on datetime, a Grid Square code, and keywords to retrieve data from the Web API

- Calculate the frequency distribution of the Grid Square code from the item search key, and the frequency distribution in terms of the Grid Square code to the word
- Calculate the similarity measure ρ in terms of the Grid Square code for a given word by Jensen-Shannon divergence.

We calculated the frequency of the Grid Square codes for several keywords and show it as a Grid Square map. Figure 8.10 shows Grid Square statistics for including Keikyu railway, Tokyu railway, Tokaido railway, and Yamanote circle railway as a Grid Square map. The area shown by the Grid Square map for these railway line names shows almost the same location as the physical space route of the actual railway line. Thus, we can identify locations from words.

Moreover, we can estimate the similarity in terms of spatial cooccurrence between words by measuring similarity by Jensen-Shannon divergence. Table 8.5 shows an example of values for the similarity ρ between words. High value of the similarity between words means that these words are observed in close areas. The closer ρ is to 1, the more words occur in the same Grid Square. As a result, co-occurring words can be compared and classified without word classification such as general words and special words if we introduce text Grid Square analysis.

8.7 Conclusions

This chapter showed several valuable examples of Grid Square statistics. Use cases of Grid Square statistics are classified into six types: data linkage and processing, mapping, data creation on given areas, identifying practical areas, defining observation areas, and units for numerical simulation.

We showed some case studies of Grid Square statistics to aggregate spatio-temporal Grid Square statistics.

We explained several types of aggregation scenarios such as approximating regional statistics by aggregating quantitative Grid Square statistics in arbitrary regions, and producing Grid Square statistics from questionnaire surveys and image analyses.

We further showed some applications of Grid Square statistics for time series analysis of floating population and regression analysis of Grid Square statistics for Tourism statistics. We proposed an approximation method to estimate hotel occupancy rates from room opportunity data collected from a hotel booking site.

We considered spatial risk assessment of natural disasters. Specifically, we estimated physical exposure to seismic hazards and census population. It was found that regional characteristics of natural disasters can be identified based on Grid Square statistics of both seismic hazard and socioeconomic variables such as population.

We mentioned an example of Grid Square statistics of SDG indicators (SDG 11.3.1 indicator) computed from Grid Square statistics of census population and satellite gridded land cover data. The indicator aggregated in the administrative division is also approximately computed from Grid Square statistics.

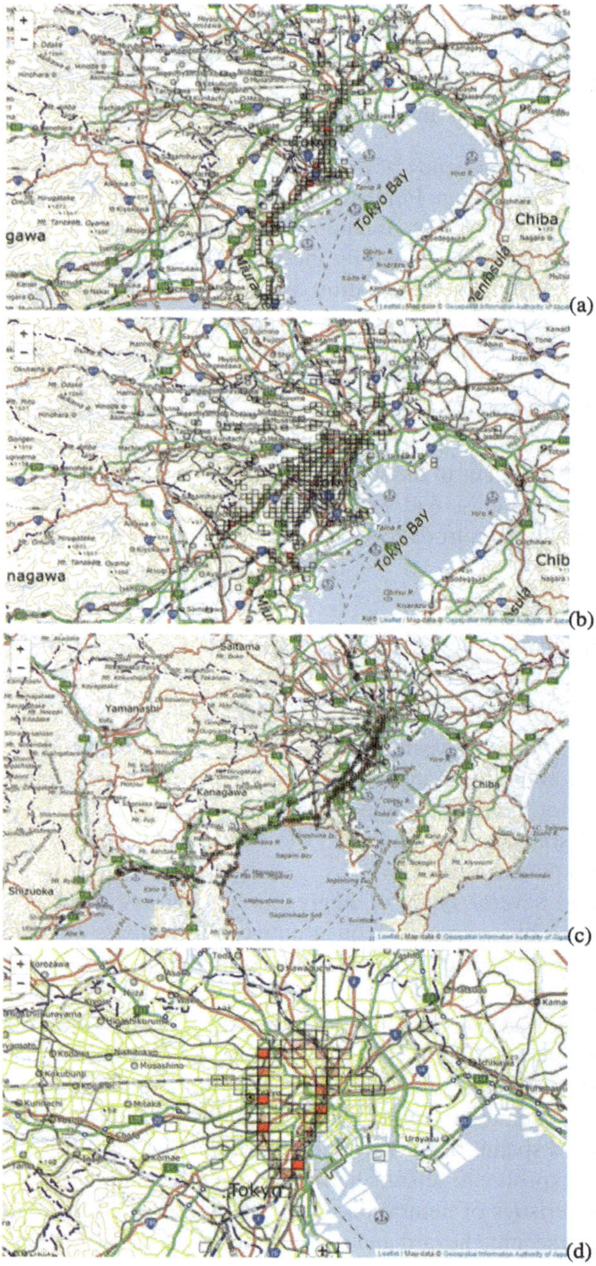

Fig. 8.10 Grid Square maps of word frequency for **a** Keikyu railway, **b** Tokyu railway, **c** Tokaido JR railway, and **d** Yamanote JR railway

Table 8.5 Examples of similarity between words in terms of Grid Squares

Word1	Word2	Similarity (ρ)
Keikyu stake	Wings Kurihama	0.809125
Keikyu stake	Chikin	0.001784
Yokosuka city	Haifuri	0.286398
Keikyu tsuna	Kimen	0.516792

We explained a method to produce Grid Square statistics about word frequency from text data with geographical information. The text Grid Square analysis helps search and understand regional characteristics from natural language. We mentioned some examples of text Grid Square statistics, including famous constructions. We introduced bipartite graph representation between words and Grid Squares to analyze the similarity of words and Grid Square and proposed a method to use Jensen-Shannon divergence as a similarity index.

Grid Square statistics can be produced from various data types such as point data, polygons, text strings, and gridded data. They are useful to capture, search, and characterize regional characteristics from various aspects such as spatio-temporal axis and natural language text. These examples can be implemented as Web data applications of Grid Square statistics in the Platform shown in Chaps. 5 and 6.

The sustainability of global systems needs to understand several types of KPIs. The international systems' scale is more extensive than our human ability. Thus, we require highly distributed and autonomous platforms to acquire, collect, process, analyze, and interpret data from global systems.

References

1. E.J. Fortuny, D. Martens, F. Provost, Predictive modeling with big data: is bigger really better? Big Data **3**(4), 215–226 (2013)
2. I. Lee, Big data: dimensions, evolution, impacts, and challenges. Bus. Horiz. **60**(3), 293–303 (2017)
3. R. Sabharwal, S. Miah, A new theoretical understanding of big data analytics capabilities in organizations: a thematic analysis. J. Big Data **8** (2021). https://doi.org/10.1186/s40537-021-00543-6
4. M. Favaretto, E. De Clercq, C.O. Schneble, B.S. Elger, What is your definition of Big Data? Researchers' understanding of the phenomenon of the decade. PLoS ONE **15**(2), e0228987 (2020). https://doi.org/10.1371/journal.pone.0228987
5. Tableau, What Is Big Data? Big Data Explained. https://www.tableau.com/learn/articles/what-is-big-data. Accessed 15 Dec 2023
6. IBM, Big Data analytics. https://www.ibm.com/analytics/big-data-analytics. Accessed 15 Dec 2023
7. M. Beyer, The importance of 'Big Data': a definition. Gartner **21** (2012). https://www.gartner.com/en/documents/2057415. Accessed 15 Dec 2023
8. S. Bigelow, "Big Data" TechTarget (2022). https://www.techtarget.com/searchdatamanagement/definition/big-data

9. SAS, "Big Data: What it is and why it matters. https://www.sas.com/en_us/insights/big-data/what-is-big-data.html. Accessed 14 Dec 2023
10. Grid Square Statistics, Statistics Bureau of Japan. https://www.stat.go.jp/english/data/mesh/01.html. Accessed 1 Jan 2024
11. Statistical GIS, Statistics Bureau, Ministry of Internal Affairs and Communications. [Online]. Available: https://www.e-stat.go.jp/gis. Accessed 10 Jan 2024
12. Population Projections for individual 1km mesh (National Spatial Planning and Regional Policy Bureau estimates for 2017). Ministry of Land, Infrastructure, Transport, and Tourism in Japan [Online]. Available: http://nlftp.mlit.go.jp/ksj/old/meta/mesh10000.html. Accessed 28 April 2018
13. Docomo Insight Marketing INC. https://www.dcm-im.com, (in Japanese). Accessed 25 Dec 2023
14. Agoop Corp. https://agoop.co.jp, (in Japanese). Accessed 25 Dec 2023
15. ZENRIN DataCom. https://www.zenrin-datacom.net/solution/congestion, (in Japanese). Accessed 25 Dec 2023
16. KDDI Location Analyzer. https://k-locationanalyzer.com/en/. Accessed Dec 2023
17. Kokusai Kogyo Co., Ltd., Wi-Fi population statistics. https://www.kkc.co.jp/service/lp/6891/, (in Japanese). Accessed 31 Jan 2024
18. Research Institute for World Grid Squares, 3rd Level Grid Square Data for SDG 11.3.1 (Ratio of land consumption rate to population growth rate). https://www.fttsus.org/worldgrids/en/sdg_11_3_1_en/. Accessed 10 Jan 2024
19. US Census, Gridded Population Mapping (Demobase) 1998-present. https://www.census.gov/geographies/mapping-files/time-series/demo/international-programs/demobase.html. Accessed 28 Jan 2024
20. X. Wang, X. Meng, Y. Long, Projecting 1 km-grid population distributions from 2020 to 2100 globally under shared socioeconomic pathways. Sci. Data **9**, 563 (2002). https://doi.org/10.1038/s41597-022-01675-x
21. F. Schug, D. Frantz, S. van der Linden, P. Hostert, Gridded population mapping for Germany based on building density, height and type from Earth Observation data using census disaggregation and bottom-up estimates. PLoS ONE **16**(3), e0249044 (2021)
22. K. Neki, Study on bhutan census grid square data creatin. Bhutan J. Res. Dev. 10–28 (2021)
23. M. Terada, T. Nagata, M. Kobayashi, population estimation technology for mobile spatial statistics. NTT DOCOMO Tech. J. 14(3), 10–15 (2013). https://www.docomo.ne.jp/english/binary/pdf/corporate/technology/rd/technical_journal/bn/vol14_3/vol14_3_010en.pdf. Accessed 14 Jan 2024
24. C. Bergroth, O. Järv, H. Tenkanen, M. Manninen, T. Toivonen, A 24 h population distribution dataset based on mobile phone data from Helsinki Metropolitan Area, Finland. Sci. Data **9**, 39 (2022). https://doi.org/10.1038/s41597-021-01113-4
25. A.-H. Sato, Microdata analysis of the accommodation survey in Japanese tourism statistics, in *2015 IEEE International Conference on Big Data (Big Data)* (2015), pp. 2700–2708
26. Japanese Tourism Statistics. https://statistics.jnto.go.jp/en/. Accessed 30 Jan 2024
27. Jalan Web Service. Available: https://www.jalan.net/jw/jwp0000/jww0001.do. Accessed 28 April 2018
28. A.-H. Sato, T. Watanabe, Measuring activities and values of industrial clusters based on job opportunity data collected from an internet Japanese job matching site, in *2016 IEEE International Conference on Big Data (Big Data)* (2016), pp. 2199–2208
29. A.-H. Sato, C. Shimizu, T. Mizuno, T. Ohnishi, T. Watanabe, Relationship between job opportunities and economic environments measured from data in internet job searching sites. Procedia Comput. Sci. **60**, 1255–1262 (2015)
30. Japan Seismic Hazard Information Station. National Research Institute for Earth Science and Disaster Resilience [Online]. Available: http://www.j-shis.bosai.go.jp/. Accessed 20 Jan 2024
31. Anticipated inundation zone data in 2012. Ministry of Land, Infrastructure, Transport, and Tourism in Japan [Online]. Available: http://nlftp.mlit.go.jp/ksj/gml/datalist/KsjTmplt-A31.html. Accessed 28 April 2018

32. A.-H. Sato, H. Sawai, Geographical risk assessment from tsunami run-up events based on socioeconomic-environmental data and its application to Japanese air transportation. Procedia CIRP **19**, 27–32 (2014)
33. National Research Institute for Earth Science and Disaster Resilience (NIED). [Online]. Available: http://www.bosai.go.jp/e/. Accessed 15 Dec 2016
34. 2030 Agenda for Sustainable Development. https://sustainabledevelopment.un.org/post2015/transformingourworld. Accessed 5 Jan 2023
35. Metadata on SDGs Indicator 11.3.1, UN-Habitat. https://unhabitat.org/sites/default/files/2020/07/metadata_on_sdg_indicator_11.3.1.pdf. Accessed 10 Jan 2024
36. 3rd Level Grid Square Data for SDG 11.3.1 (Ratio of land consumption rate to population growth rate), Research Insitite for World Grid Squares. https://www.fttsus.org/worldgrids/en/sdg_11_3_1_en/. Accessed 31 Jan 2024
37. JAXA HRLULC Map: High-Resolution Land Use and Land Cover Map. https://www.eorc.jaxa.jp/ALOS/en/index_e.htm. Accessed 30 Jan 2024
38. J. Lin, Divergence measures based on the Shannon entropy. IEEE Trans. Inf. Theory **37**(1), 145–151 (1991)

Index

A.-H. Sato and H. Tsubaki, *Evaluation Platform of Sustainability for Global Systems*, https://doi.org/10.1007/978-981-97-2296-9

SPRINGER NATURE

GPSR Compliance

The European Union's (EU) General Product Safety Regulation (GPSR) is a set of rules that requires consumer products to be safe and our obligations to ensure this.

If you have any concerns about our products, you can contact us on ProductSafety@springernature.com

In case Publisher is established outside the EU, the EU authorized representative is:

Springer Nature Customer Service Center GmbH
Europaplatz 3
69115 Heidelberg, Germany

The manufacturer's authorised representative in the EU is Springer Nature Customer Service Centre GmbH, Europaplatz 3, 69115 Heidelberg, Germany. If you have any concerns regarding our products, please contact ProductSafety@springernature.com

Printed and bound by CPI Group (UK) Ltd, Croydon, CR0 4YY
01/10/2026
02225875-0009